DR KARL'S
LITTLE BOOK OF
CLIMATE
CHANGE
SCIENCE

Dr Karl Kruszelnicki

ABC
BOOKS

PRE-EMPTIVE ERRATUM

We have gone to a lot of trouble to fact-check as thoroughly as we could. If any mistakes have slipped through the cracks, please let us know at mistakes@drkarl.com and we will fix them next time around.

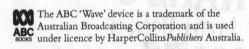 The ABC 'Wave' device is a trademark of the Australian Broadcasting Corporation and is used under licence by HarperCollins*Publishers* Australia.

HarperCollins*Publishers*
Australia • Brazil • Canada • France • Germany • Holland • Hungary
India • Italy • Japan • Mexico • New Zealand • Poland • Spain • Sweden
Switzerland • United Kingdom • United States of America

First published in Australia in 2021
by HarperCollins*Publishers* Australia Pty Limited
Level 13, 201 Elizabeth Street, Sydney NSW 2000
ABN 36 009 913 517
harpercollins.com.au

A catalogue record for this book is available from the National Library of Australia.

ISBN 978 0 7333 4129 8 (paperback)
ISBN 978 1 4607 1303 7 (ebook)

Cover and text design by Lisa Reidy
Cover and author photo by Steve Baccon
Typeset in Chaparral Pro by Kelli Lonergan
Printed and bound in Australia by McPherson's Printing Group
The papers used by HarperCollins in the manufacture of this book are a natural, recyclable product made from wood grown in sustainable plantation forests. The fibre source and manufacturing processes meet recognised international environmental standards, and carry certification.

CONTENTS

INTRODUCTION

WHY I'M STILL WRITING ABOUT CLIMATE CHANGE

It was back in 1981 that I did my first story on Climate Change for Triple J youth radio. Back then, the evidence had been building up for more than 20 years. About a decade later, in 1990, the climate scientists finally agreed that Climate Change was happening — and that we humans were causing it.

Forty years after my first story, I'm still writing about it, and now I've decided to write a book. Here's why.

THIRTY YEARS IS TOO LONG

It took only two years to create laws banning CFCs, once we'd proven that they were destroying our protective ozone layer. So why are we still arguing about Climate Change 30 years after we already proved it was caused by human emissions of Greenhouse Gases?

Because Big Fossil Fuel started political lobbying and a massive disinformation campaign that was already spending a billion dollars a year by 2013. They were even more successful at spreading confusion than Big Tobacco and Big Alcohol.

They started with, 'Global warming isn't happening.'

This turned into, 'Okay, global warming is happening, but it's entirely caused by nature, not by human activity.'

The third version was, 'Okay, we are causing Climate Change, but it's actually going to be good for people and the planet.'

And now the current line is, 'Okay, we are causing Climate Change, which is expensive and bad – but it's too late to fix it.'

Thanks to Big Fossil Fuel's very successful lobbying, the human race wasted three precious decades. If we hadn't, we could have avoided the five major coral bleachings of the Great Barrier Reef.

BYPASS THE SPIN

When it comes to discussing the science of Climate Change, there's too much heat and not enough light. In Australia, many sections of the media (radio, TV, print and interwebs) give more coverage to lies or half-truths about Climate Change than to the full truth. They also give the impression that the majority of the community actively denies it exists.

However, only 3% of the population globally (and only 8% of the Australian public) actively denies the science of Climate Change. And the figure for the experts actually working in the field (the climatologists) is about 0%.

So, despite the firehose of disinformation, clearly the public is not as gullible as some media and politicians would like to believe.

In this book, I shall bypass the spin and stick to the facts.

SCIENTISTS ARE JUST DOING THEIR JOB

I feel sorry for the climate scientists. They have been getting huge amounts of heat for simply doing their job in reporting the data and telling us the facts. Death threats, eggings, public humiliation and getting fired (okay, 'made redundant', if you prefer a euphemism) is not part of anybody's job description. I hope this book will make their lives easier by supporting them and their research. After all, if it wasn't for scientists, we'd be more ignorant.

IT'S NOT TOO LATE – AND THERE'S MORE GOOD NEWS!

Best of all, it's not too late!

We *can* still fix Climate Change (hooray!). We can stop putting Greenhouse Gases from burning fossil fuels *into* the atmosphere (and still easily get all the energy we need to run our civilisation). Even better, we can reverse Climate Change – by removing Greenhouse Gases *from* the atmosphere.

The advantages include much cheaper energy and cleaner air (and fewer deaths). Renewable energy industries also create many more jobs than fossil fuel industries – jobs that are mostly local.

FIVE THINGS YOU'LL FIND OUT MORE ABOUT

Here's the executive summary:

1. There's been a three-decades-long well-funded disinformation campaign to cover up the facts of Climate Change. Unfortunately for the planet, it's been very successful.

2. Why did we start burning carbon-based fossil fuels? Easy answer – it's because they are loaded with energy. One barrel of oil (at a cost of A$50) carries the energy of a labourer working for ten years (at a cost of A$500,000).

3. The 'warmth' of Global Warming doesn't come directly from the heat of combustion released by burning fossil fuel. Burning releases carbon dioxide, which traps extra heat from the Sun (as this heat goes through its regular cycle of entering and then leaving the biosphere). How much heat is trapped? About 400,000 Hiroshima atom bombs each day!

4. Climate Change affects us and the planet in lots of ways. One effect is that we've actually tipped the Earth off its axis (by a tiny amount,

but still ...). And there are other effects – droughts, bushfires, migrating diseases, hurricanes that cause more economic damage to a country than the value of its Gross Domestic Product – and many more.

5.   The only thing stopping us from fixing Climate Change is point 1 above – the billion dollars (and more) spent by the fossil fuel industry every year on deliberate disinformation. We also don't need any amazing new breakthroughs in science or technology to fix Climate Change (although they will certainly happen – and help). We just need the political will to start using and developing our existing science and technologies.

Today's generation is the smartest ever in recorded humanity, *and* we are living in the most peaceful time ever in recorded history. So the odds of fixing Climate Change are excellent.

I'm an eternal optimist. With a combination of Intelligence, Peace and Knowledge we can fix the problem and improve the lives of current, and future, generations. We've just got to get started – but not next century, or next decade, or next year. We've got to get started *now*!

1
HOW LONG HAS THIS BEEN GOING ON?

A snapshot of the science – and how it was covered up

Scientists have been investigating the temperature of the Earth's atmosphere for centuries. But for a few decades in the second half of the 20th century, fossil fuel companies got in on the action. At first, they did real science – but then they shifted to chicanery.

As far back as the mid-1600s, the 'Medici Network' (which sounds a bit like an organised crime syndicate but was actually the world's first weather service) was already taking continuous measurements of air temperature across Europe.

The experimental basis of Global Warming was established in 1856, when the American scientist Eunice Newton Foote measured how different gases absorbed the heat of the Sun. As a result, she correctly predicted that extra carbon dioxide in the atmosphere would increase air temperature.

In the 20th century, surprisingly, fossil fuel companies carried out much of the research into Climate Change, beginning in the mid-1970s. In 1982, they predicted, with remarkable accuracy, what the carbon dioxide levels and air temperatures would be in 2020 if the current trend in burning fossil fuels continued. Back then, they accepted the science of Climate Change and, as good corporate citizens, they rejected projects that would release too much carbon dioxide into the atmosphere.

Around 1990, however, Big Fossil Fuel did a U-turn and began a very active and massively well-funded cover-up of the real science.

And here we are at the start of the third decade of the 21st century. Let's dive into the details of how we got to our current state of Global Warming by looking at how the science was done – and how it was covered up.

THE SCIENCE

1654: Medici Meteorological Network, 'the world's first weather service', measures air temperature continuously for 16 years

Probably the earliest known temperature measurements ran from 1654 to 1670. The so-called 'Medici Network' took temperature measurements every 3 to 4 hours at 11 locations. The two main stations were in Florence (at sea level) and Vallombrosa (1 km altitude) in Italy, with other stations scattered across Europe (including Paris, Warsaw and Innsbruck). The enterprise was funded and supported by the Medici family, which supported science in the same way it supported the arts. After all, at that time, it was not known if ice always melted at the same temperature, or what the range of air temperatures was in different countries, for example.

The stations all used identical thermometers (the newly invented Little Florentine Thermometer) mounted in a similar manner – pretty smart.

1824: Fourier figures out that the atmosphere traps heat

In 1824, Joseph Fourier, a French physicist and mathematician, published his deep thoughts about Earth's atmosphere. Why was Earth's temperature what it was, and not hotter or colder?

There were, Fourier realised, three potential sources of heat for the Earth: the Sun, Earth's core and 'space' (the rest of the Universe).

He also realised there was a fourth factor: Earth's atmosphere. Somehow, this 'trapped' some of the Sun's heat, preventing it from leaving, and so warming the Earth's surface. Without an atmosphere, he observed, the Earth would be much colder. Fourier was one of the first to recognise what would be later called the 'Greenhouse Effect'.

Fourier's work was purely theoretical, however – he didn't carry out any experiments.

1838: Pouillet measures the power of the Sun

In 1838, the French physicist Professor Claude Pouillet measured the solar constant (the average amount of power per square metre the Sun delivers to the Earth, at the upper edge of the atmosphere). Using equipment (called a pyrheliometer) he built himself, he estimated that the Sun delivers 1,228

Watts per square metre, or Watts/m² (remarkably close to today's more accurate estimate of 1,361 Watts/m²).

Pouillet was a gifted experimenter, but he didn't identify any atmospheric gases that could trap the sun's heat.

1856: Foote shows that CO_2 gets hotter than other gases

In 1856, Eunice Newton Foote, an American scientist, inventor and women's rights campaigner, did the first experiment to identify which gases could force the atmosphere to heat up. Foote filled glass cylinders about 75 cm long and 10 cm in diameter with different gases, such as hydrogen, oxygen, carbon dioxide and so on. She left them in direct sunlight to find out how much each gas heated up.

She discovered that hydrogen heated up the least, while carbon dioxide became the hottest. Then Foote made the chilling prediction that 'an atmosphere of that gas [carbon dioxide] would give to our Earth a high temperature'.

1860: Tyndall discovers that CO_2 in the air traps heat

Irish physicist John Tyndall took the next step with a series of experiments around 1860.

Tyndall had the dual advantages of being a man, and of having easy access to the considerable resources of England's Royal Institution (an organisation dedicated to scientific research and education). His experiments measured that carbon dioxide could trap nearly 1,000 times as much heat as dry air.

Tyndall summarised the physics behind Global Warming accurately: 'Thus the atmosphere admits of the entrance of the solar heat; but checks its exit, and the result is a tendency to accumulate heat at the surface of the planet'.

So, all the way back in 1860, we discovered that the atmosphere will easily let *in* the Sun's heat – but will stop most of it from leaving.

1894: Högbom explains the Carbon Cycle

By 1894, scientists were very confident that carbon dioxide was deeply implicated in helping to set our atmosphere's temperature. But back then, they had only a very poor understanding of how carbon atoms moved around in the biosphere (for example, how they move from the atmosphere into a tomato, into you, then out of your mouth and into the ocean, where they dissolve and then become part of a coral, then into a fish, etc.). This movement of carbon atoms into (and out of) natural and human sources (and sinks) is called the Carbon Cycle.

The Swedish geologist Arvid Högbom did much of the early work on the Carbon Cycle in the late 1800s.

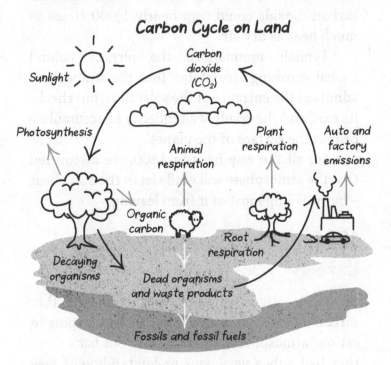

Carbon Cycle on Land

(Based on illustration at onlinesciencenotes.com/the-carbon-cycle/)

1896: Arrhenius concludes that humans are heating the planet

Arvid Högbom and Svante Arrhenius were colleagues. Högbom's ground-breaking work on the

Carbon Cycle was essential for Arrhenius, who tried to predict how changing carbon dioxide levels in the atmosphere would affect the Earth's temperature.

In 1896, after doing tens of thousands of calculations – all by hand, without a calculator – Arrhenius concluded that human activity was making the Earth's atmosphere warmer.

In 1906, he fine-tuned his work, predicting: 'Any doubling of the percentage of carbon dioxide in the air would raise the temperature of the Earth's surface by 4°C.' This was impressively close to current predictions.

But Arrhenius missed the mark when he predicted it would take humanity 3,000 years to increase atmospheric carbon dioxide levels by 50%.

No.

In the years since 1900 (just 120 years), we have already raised carbon dioxide levels by 40%.

1938: Callendar links fossil fuels to rising temperatures

In 1938, Guy Stewart Callendar, an English steam engineer, inventor, and scientist, showed the link between temperature and carbon dioxide. After years of effort, he finally collected enough data from around the planet to demonstrate that rising carbon dioxide levels corresponded with increased burning of fossil fuels – and that both were linked to rising atmospheric temperatures.

Callendar proposed an organised research program to continuously measure carbon dioxide levels in the atmosphere.

1958: Keeling shows that CO_2 levels are rising

In 1958, the American scientist Charles David Keeling set up a station at Mauna Loa Observatory, Hawaii, some 3.4 km above sea level, specifically to measure carbon dioxide levels in the atmosphere.

Monthly mean CO_2 concentration

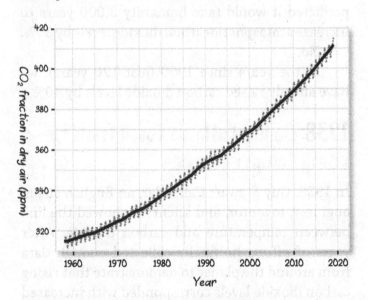

Atmospheric CO_2 concentrations measured at Mauna Loa Observatory: The Keeling Curve
(Data: Dr Pieter Tans, NOAA/ESRL, and Dr Ralph Keeling, Scripps Institution of Oceanography)

Three years later, in 1961, he had enough data points to show that the level of carbon dioxide in the atmosphere was steadily rising. In just those three years, it rose from 315.34 parts per million (ppm) to 317.64 ppm – about a 0.77 ppm rise in each year. Since then, the rate of increase has risen to the current 2.46 ppm per year, with carbon dioxide levels reaching 415 ppm around the start of 2021.

1973: Insurance companies are alert and quite alarmed

Munich Re, the mammoth reinsurance company, proudly bills itself as 'the first alerter to global warming'. (An 'insurance' company will insure, for example, your car. But what if a massive hailstorm damages 100,000 cars? To protect itself from that potential massive payout, your insurance company will take out insurance with a much bigger company – a 'reinsurance' company, such as Munich Re.) Back in 1973, the company drew 'attention to man-made climate change and its effects'. Munich Re had noted that flood damage was increasing. It predicted: 'In the long-term, global warming will lead to a further increase in weather-related natural catastrophes, the financial impact of which will have to be borne by insurers and the public.'

Climate Change will cost you

You don't find many climate-change deniers in the reinsurance business. Insurance companies are very keen to not lose money, and reinsurers have seen the costs of weather-related claims double every decade since 1980.

Of course, because none of the costs of Climate Change are carried by the fossil fuel companies that cause most of it, insurers pass these costs directly to the public – you and me – by charging higher premiums.

It's hard to work out the exact cost of Climate Change. It will reach into many aspects of our world – agriculture, coastal storms, energy supplies, crime, etc. Usually, costs to the environment and the cost of people's health are not even factored in.

Currently, the annual financial cost of Climate Change is thought to be at least 1% of world gross domestic product (GDP), which sits at about US$85 trillion. That's quite separate from annual fossil fuel subsidies of over US$5 trillion. One recent estimate is that, by 2070, the annual financial costs of Climate Change will be about 15% of world GDP.

1977: Big Fossil Fuel takes the science to the next level

This next part of the story may surprise you. In 1977, Exxon Corporation's senior scientist, James Black, told the company's management committee: 'There is general scientific agreement that ... mankind is influencing the global climate through carbon dioxide release from the burning of fossil fuels'. James Black also wrote: 'Some countries would benefit, but others would have their agricultural output reduced or destroyed.'

At various times, Exxon has been one of the world's largest fossil fuel companies, so it had always planned for the long term. Realising they urgently needed more data about the impact of carbon dioxide on Earth's temperature, they embarked on the research, boots and all.

At that stage, nobody had the exact figures on how much carbon dioxide stayed in the air, went into the ocean, was reabsorbed into plants, broke down or was incorporated into other chemicals. In order to get hard data on just one aspect of the movement of carbon dioxide, within three months (amazingly quickly), Exxon had modified their largest supertanker, the *Esso Atlantic*, to monitor carbon dioxide levels in the oceans for the next three years, to see how much was dissolving there. It cost them about US$1 million.

1980s: Good science and good corporate citizens

By 1980, Exxon had assembled teams of the world's most sophisticated climate modellers to investigate fundamental questions about carbon dioxide build-up. Exxon teamed up with Amoco, Gulf Oil, Mobil, Phillips, Shell, Standard Oil of California, Sunoco, Texaco and Sohio to share their research into climate science. This consortium of genuine scientists was first called 'The CO_2 and Climate Task Force', but was soon changed to 'The Climate and Energy Task Force'.

In 1982, their theoretical scientists reported that, in the longer term, the doubling of the carbon dioxide level (up to 560 ppm) would produce an average atmospheric Global Warming of about 3°C.

Other scientists predicted that, assuming the rate of increasing carbon dioxide production continued, the atmospheric temperature would rise by 1°C in the medium term (by 2020), and that atmospheric carbon dioxide would rise to about 410 ppm.

These predictions, made 38 years earlier, were almost spot on!

In the 1980s, Exxon decided not to develop one of the world's largest gas fields – Natuna, off the north coast of Indonesia, in the South China Sea –

because, besides containing about 30% methane (natural gas), the gas contained some 70% carbon dioxide. All that carbon dioxide would have to be released into the air, which would make the gas field the largest single-point generator of carbon dioxide on Earth (about 1% of global emissions).

So, at this time, Exxon were behaving like good corporate citizens. The company even employed Dr James Hansen, a leading climate expert with NASA, to help them understand what was then still called the Greenhouse Effect. (Hansen later helped popularise the term 'Global Warming' when he used it in June 1988 in his testimony to the US Congress.)

From hero to hated

Later, Dr James Hansen was unfairly vilified by Exxon's climate denialist campaign. He went from being one of their respected and beloved experts to somebody they hated, after he told the US Congress in 1988 that the Greenhouse Effect (Global Warming) had definitely arrived. His message hadn't changed – but Exxon's attitude had.

1998: Climatologists publish their 'Hockey Stick Graph', which becomes the most important graph of climate science

In 1998, after years of complex research, climatologists Michael Mann and Raymond Bradley and meso-climatologist and tree-ring specialist Malcolm Hughes together published a paper in *Nature* about global temperature patterns from the year 1400 to 1998. The paper contained a graph that's since become known as the Hockey Stick Graph. It shows a long and very slow cooling trend over nine centuries of around 0.02°C per century (the shaft of the ice-hockey stick), and then a sudden heating beginning in the late 19th century and continuing into the 20th century of about 0.8°C per century (the sharply curved blade of the stick). (Australia's average temperature has increased by 1.44°C since 1910, when national records began.)

To produce the graph and their paper, Drs Mann, Bradley and Hughes first had to collect data from other scientists. But given we've had global temperature-measuring networks for only the past century, how could scientists estimate temperatures before, say, 1900?

The answer is they use imperfect natural thermometers (called proxies). Proxies include ice cores, coral, plant pollen, insect and animal remains, tree rings, lake sediments and more. They then use various scientific techniques to estimate the ambient temperature at the time these proxies formed. These techniques include the ratio of heavier and lighter isotopes of atoms such as oxygen and hydrogen (which make up water). For example, when water evaporates naturally, the molecules made of lighter isotopes leave the body of liquid before the molecules made of the heavier isotopes, and this behaviour changes with increasing ambient temperature.

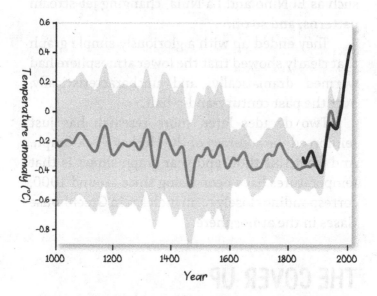

The light curve is Mann, Bradley and Hughes' original Hockey Stick Graph from a 1999 paper, with its uncertainty range in a lighter colour. The dark curve shows the temperature differences, between 1850 and 2013, between actual measured temperatures and the long-term average.
(Klaus Bitterman/OTRS/Wikimedia Commons)

As you can imagine, scientists had gone to great lengths to gather the data from ocean coral reefs, ancient dry lakes now overgrown by jungle, ice cores from not just the Arctic and Antarctic regions but also remote mountain tops, etc.

Then Mann, Bradley and Hughes had to devise new mathematical techniques, so they could compensate for the average global temperature being complicated by natural background changes, such as El Niño and La Niña, changing jet-stream patterns, and so on.

They ended up with a gloriously simple graph that clearly showed that the lower atmosphere had warmed dramatically, and uncharacteristically, over the past century-and-a-half.

Two decades later, more research has just reinforced the accuracy of the Hockey Stick Graph. And what this thousand-year graph shows is that temperatures have been rising since around 1900, corresponding closely to an increase in Greenhouse Gases in the atmosphere.

THE COVER-UP

1990s: Big Fossil Fuel denies the science of Climate Change

By around 1990, Big Fossil Fuel had changed its goals with regard to climate science. The transparent research that fossil fuel companies had initially conducted came to an abrupt end.

Instead, they formed the 'Global Climate Coalition'. This was strictly a political lobby group that worked behind the scenes. Its objective was to deny and discredit Global Warming. This group worked relentlessly and secretively to keep fossil fuels in the market, and successfully blocked any government efforts to reduce the burning of fossil fuels.

From stating, in 1978, that there was 'general scientific agreement' about Climate Change, by 1997, Big Fossil Fuel's position shifted to the false claim that 'scientific evidence is inconclusive'.

Back in 1978, Exxon Senior Scientist, James Black had said that humanity had 'a time window of five to ten years before the need for hard decisions' with regards to ditching fossil fuels. By 1997, however, Lee Raymond, the Chairman and CEO of Exxon, said that nothing would change in the climate if 'policies are enacted now or 20 years from now'.

Why the reversal?
The Big Fossil Fuel companies knew that this current episode of Climate Change was real and was caused by burning fossil fuels. It's possible they realised they had two options:

Option A. They could continue to sell 'energy', but gradually move out of fossil fuels and into other sources of energy. After all, the world still needed energy (heat, electricity, transport, etc). But this was new territory. Most companies would do well, but some might not. At the time, this probably seemed like a riskier option.

Option B. They could do Business As Usual (BAU) and keep selling and burning fossil fuels. This would be, in the short term of a few decades, a lower risk option. But this pathway would require high-level government lobbying (via the Global Climate Coalition), as well as a well-funded disinformation campaign. Luckily, when it came to BAU and disinformation, Big Fossil Fuel had the successful examples of Big Tobacco and Big Alcohol to follow. (After all, Big Tobacco and Big Alcohol are still doing very well.)

Big Fossil Fuel took the BAU pathway.

The lobbying by Big Fossil Fuel was so powerful that, in 2001, the newly installed President of the USA, George Bush, appointed Philip A. Cooney from the American Petroleum Institute, as the new Chief of Staff for the Council on Environmental Quality – the office at the White House that directed its policy on climate. Now the fox was 'guarding' the chicken coop!

In 2005, Cooney resigned to work for ExxonMobil. Under George Bush, the USA also pulled out of the Kyoto Protocol, an international treaty with the aim of limiting global Greenhouse Gas emissions.

But after a decade or so, scientists took Big Fossil Fuel to task.

1998 to early 2000s: Big Fossil Fuel tries to discredit the Hockey Stick Graph

After the simple and precise Hockey Stick Graph was published in 1998, Big Fossil Fuel funded huge, multi-year-long disinformation campaigns to try to discredit it.

Climate scientists received death threats. Dr Mann had the phone number of the local police station taped to his fridge, so his family could quickly call for help.

On the political side, Senator James Inhofe of Oklahoma attacked Dr Mann's work on the floor of the US Senate, even though Senator Inhofe had no scientific training. Representative Joe Barton tried to subpoena every single email from Dr Mann's entire email history, as well as his financial history, apparently to try to find something to discredit Dr Mann. (Coincidentally, at the time, Inhofe was one of the largest recipients of oil and gas funding in the US Senate, while Barton was one of the largest recipients of oil and gas funding in the US House of Representatives.)

Many non-climatologists (such as economists or statisticians with no experience in any of the physical sciences) were funded to try to show the Hockey Stick Graph was wrong. But it wasn't. It was reliable and accurate.

2006: Royal Society complains

The Royal Society is probably the world's most prestigious science institution. It had both Isaac Newton and Albert Einstein as members.

In 2006, the senior manager for policy communication at the Royal Society, Bob Ward, sent a harsh letter to Exxon. He pointedly blamed Exxon for being 'inaccurate and misleading' with regard to climate science. He also accused Exxon of funding dozens of organisations that told deliberate lies about climate science.

It took two years but, eventually, Exxon stopped funding *some* of these misleading groups named by the Royal Society.

But in other arenas, Exxon (and others) still kept funding Climate Denialism. In 2013, *The Guardian* reported that 'Conservative groups may have spent up to $1 billion a year on the effort to deny science and oppose action on climate change'.

2010: Business As Usual going well?

By 2010, Exxon was doing very well indeed. BAU seemed to be working for them.

In fact, on the same day that Tesla was launched on the US stock market (29 June 2010), Exxon was the most highly valued company on the Dow Jones Industrial Average (a stock market index that measures the stock performance of the 30 wealthiest companies in the USA).

2013: Fossil fuel subsidies – 8% of all government revenue, worldwide

Certainly, Exxon's government-level lobbying reaped them huge profits, but it wasn't so good for the rest of us.

In 2013, the International Monetary Fund (IMF) released a report called *Energy Subsidy Reform*. The IMF estimated that, worldwide, government subsidies (that is, free money) for energy (petroleum products, electricity, natural gas and coal) were at least US$1.9 trillion in 2011 – about 2.5% of the entire global GDP.

These subsidies made up 8% of all revenues generated by all the governments on our planet – which is a huge amount! Out of each dollar earned by the governments (in taxes, selling export goods, etc.), eight cents are given to Big Fossil Fuel – and never has to be given back. Imagine how many schools and hospitals that would buy!

Did anyone benefit from these subsidies, except the fossil fuel companies?

No.

According to the IMF, these subsidies actually depressed economic growth.

And, of course, they were also bad for the environment, people's health and society in general.

2020: Still Business As Usual?

Remember that back in 2010, on the day Tesla (the electric vehicle and clean energy company) went public, Exxon was the world's most valuable company?

Just ten years down the track, things were very different.

Tesla's value was now more than the next nine biggest car companies in the world combined!

But Exxon had done so poorly, financially, that it no longer even appeared on the Dow Jones Industrial Average.

The old saying that a bargain with the Devil will usually end badly seems to be right in this case – but it took a long enough time.

How badly did Exxon do? Here are three examples.
Example 1. Let's suppose that, back in 2010, you'd invested $1000 in both Exxon (the world's wealthiest

company) and NextEra (an energy company with a large focus on wind and solar power).

By 2020, excluding dividends, your $1000 investment would have returned a profit of 600% on NextEra ($6000), but a loss of 25% on Exxon ($750).

Example 2. How did Exxon and NextEra do financially in the first half of 2020?

NextEra made a net profit of US$1.7 billion.

Exxon made a net loss of US$1.7 billion.

Example 3. How did their company value on the stock market compare?

Yup, NextEra was worth more than Exxon.

So, what future pathway has Exxon chosen? According to *Scientific American*: 'As other oil majors have committed to net-zero emissions goals and invested in renewables, Exxon plans to increase drilling'. Exxon seems to be going alone on this.

2021 and beyond: Fossil fuel a stranded asset

It is very clear that, worldwide, fossil fuels are becoming a 'stranded asset'. Stranded assets are 'assets that have suffered from unanticipated or premature write-downs, devaluations or conversion to liabilities', according to the Smith School of Enterprise and the Environment at the University of Oxford.

Take sails that were used on sailing ships, for example. This was an excellent industry to be in – until steam engines were installed on ships. Suddenly, 'sails' were a stranded asset – nobody wanted them anymore, and the value of sail-making companies plunged.

The same thing happened with camera film, once we could take photos with our phones.

In October 2020, the Japanese prime minister pledged that Japan would achieve net-zero carbon emissions by 2050. In 2019, Japan bought about A$31 billion of Australian energy resources – A$9.6 billion of coal, and A$21 billion of natural gas. Prakash Sharma, the Asia Pacific head of markets and transitions for Wood Mackenzie, said, 'Japan's demand for LNG [Liquified Natural Gas] and coal was already slowing considerably over the last couple of years. ... the decline will accelerate post-2030, because this is what net-zero is all about: moving away from fossil fuels, using more renewables, electricity, hydrogen.'

Australia's fossil fuel resources are rapidly turning into stranded assets. Perhaps we should invest in something more high-tech, rather than getting stuff out of the ground and selling it at dirt-cheap prices!

In a way, it's hopeful to reflect that Big Fossil Fuel did once have an open mind and a good heart. Maybe they can return to a more genuine position – and convert oil refineries to hydrogen production plants.

Getting the facts straight – how the disinformation campaign affects Australia

The disinformation campaign by the fossil fuel industry has been remarkably effective in Australia.

First, look at natural gas (methane) and employment. A 2020 survey showed that the average Australian thinks the number of people involved in 'gas mining and exploration' make up 8.2% of the Australian workforce of 12.5 million workers. The true figure is less than 0.2%.

Second, look at the revenue the Australian federal government collects from oil and gas companies, who sell these underground resources. The same survey showed that most Australians think this revenue contributed about 10.8% of the Commonwealth budget for 2018–2019. The true figure was 0.2%.

Despite the massive decades-long disinformation campaign, however, it's reassuring that the percentage of committed climate denialists is very low at around 8% (in Australia). The reason that it's more than zero is deliberate disinformation in some news media and on social media.

Take Twitter. About one-quarter of all tweets saying that the science of Climate Change is 'flawed' are generated by 'bots' – not humans. Bots are software that can, by themselves, tweet any of a large number of messages, retweet, and even 'like' other tweets. At first glance, these tweets appear to come from people like you and me – but they don't. This 'Twitter Bot Campaign' is just another example of the well-orchestrated and sophisticated machine set up to deny climate science. The purpose of these tweets is to fool you into thinking that most people deny climate science.

It's not too late!

2
SHIMMYIN' AND A-SHAKIN'

What's a Greenhouse Gas and how does it trap heat?

Why do we talk about the Greenhouse Effect and Greenhouse Gases? Well, certain gases in the atmosphere perform a similar function to the glass in a greenhouse, which is a building with glass walls with plants growing inside. The glass walls of a greenhouse let in the heat and light from the Sun. The trapped heat then warms up the air inside the greenhouse. Thanks to transparent glass and warm air, you can grow crops in winter.

Greenhouse Gases work a bit like the glass in a greenhouse. They let the incoming heat from the Sun pass through and hit the Earth's surface – in other words, they're transparent to heat from the Sun. But they're not transparent to outgoing heat from the surface of the planet. So Greenhouse Gases trap extra heat in the atmosphere – which then warms up.

Greenhouse Gases only make up less than 1% of the atmosphere. So what makes a gas a Greenhouse Gas? The fact that it's made up of three or more atoms. This means it can move and groove and vibrate in many more ways than a gas made of just one or two atoms. In some cases, these vibrations are exactly tuned to trap the outgoing heat, and now we have a Greenhouse Gas ...

SOMETHING'S IN THE AIR

Our Earth's atmosphere is made of about 78% nitrogen, 21% oxygen, about 1% argon – and a whole smattering of other relatively uncommon molecules such as carbon dioxide, methane, neon, oxides of nitrogen, chlorofluorocarbons, etc. (The above percentages are for 'dry' air – with zero water molecules. In our atmosphere, the air's humidity, or percentage of water vapour, is around 0.4% – but it varies from 0.01% to over 4%. Humidity is higher during a monsoon in the tropics, and lower in the middle of a desert in the heat of summer.)

A tiny fraction of the gases in the atmosphere are Greenhouse Gases. Even though they're not plentiful, they've trapped enough of the Sun's heat to give us our average atmospheric temperature of about 14°C (well, that was the average temperature on Earth just before the Industrial Revolution). Without these gases, Earth would be as cold as our Moon, about -23°C.

There are different Greenhouse Gases, and they all absorb different levels of heat. In 2019, the different gases absorbed the following percentages of the total heat trapped:

- carbon dioxide (CO_2, from burning, agriculture, etc.) – 66%
- methane (CH_4, from escaping natural gas, agriculture, etc.) –16%
- chlorofluorocarbons (from air conditioning, refrigeration, etc.) and other gases – 12%
- nitrous oxide (N_2O, from agriculture, industry) – 6%.

One slight complication is that, molecule for molecule, each Greenhouse Gas causes a different amount of heating of the atmosphere. For example, one molecule of methane (CH_4) traps much more heat than one molecule of carbon dioxide (CO_2). (Luckily, there's not a lot of methane in the atmosphere – at the moment.) On the other hand, methane doesn't survive in the atmosphere as long as carbon dioxide does.

When you do the swings-and-roundabouts, methane is considered to be 84 times more effective at trapping heat than carbon dioxide over a 20-year period, and 28 times more effective over a 100-year period.

GREENHOUSE GASES SHIMMY AND SHAKE

So what is it about Greenhouse Gases that causes them to absorb heat? Well, it's to do with the three or more atoms they have.

Gases with just one or two atoms don't absorb heat very well, so they have very weak (essentially zero) Greenhouse Gas activity. This is because gases with so few atoms can vibrate in only a few ways, so the chances that one of these vibrational modes could absorb heat are very low. These gases include argon (Ar) with just one atom, or nitrogen (N_2) or oxygen (O_2) with two atoms each.

But gases with three or more atoms absorb more heat. Why?

Well, because they can resonate with heat energy in many more different ways and absorb them.

Greenhouse Gases can absorb heat in the same way a tuning fork can absorb sound waves. And when the tuning fork absorbs that sound energy, it stops it from going where it was heading.

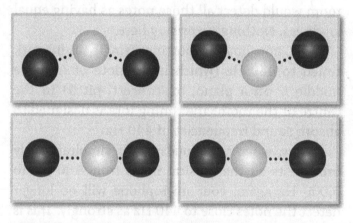

Here's a ball-and-stick representation of a molecule with three atoms – two black atoms at the end of each arm, and one single grey atom in between. There are lots of ways the molecules can shimmy and shake. For example, the angle between the two arms can change, one arm can get longer while the other arm can shorten, or each of the three atoms can rotate or spin independently. These are just some of the ways Greenhouse Gas molecules can 'vibrate'.

Think of it this way. Imagine you're in a room with a loudspeaker on one side and a microphone on the other.

Now, the sound frequencies (also called pitch) of the human voice typically range from 100 Hz to about 5,000 Hz (with consonants at the higher end and vowels at the lower end). If you could get your loudspeaker to simultaneously emit all of the notes from about 100 Hz to 5,000 Hz at equal loudness, the microphone on the other side of the room would detect all those notes as having equal loudness. Nothing surprising here.

OK, now imagine putting a single tuning fork tuned to 440 Hz (which is the note of A above middle C on a piano, or 'concert' pitch) in the centre of the room. The tuning fork is designed to absorb sound frequencies of 440 Hz.

Now, when your loudspeaker simultaneously broadcasts all the musical notes from 100 Hz to 5,000 Hz again, your microphone will no longer detect the notes close to 440 Hz as strongly. This is because the sound waves in the room that happen to be close to 440 Hz 'resonate' with the tuning fork. Some of the sound energy at about 440 Hz, which would have gone to your microphone, instead turns into mechanical energy – and forces the arms of the tuning fork to vibrate. That's why less of this particular sound energy (around 440 Hz) does not get to the microphone.

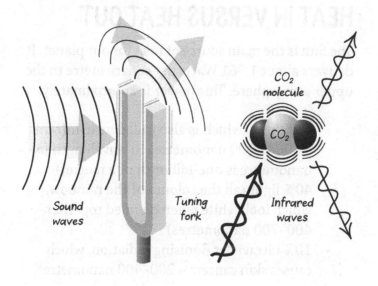

Sound waves

Tuning fork

CO_2 molecule

CO_2

Infrared waves

(Based on illustration by JG/Skeptical Science)

So, just as the tuning fork vibrates, so do gases with three or more atoms. If such a gas happens to vibrate in sympathy with any electromagnetic radiation passing through, it can absorb some of the energy present in that radiation.

This is what Greenhouse Gases in the atmosphere do. They resonate with the heat energy leaving the surface of the Earth and trap it. They prevent some of that heat energy from leaving the atmosphere.

HEAT IN VERSUS HEAT OUT

The Sun is the main source of heat for our planet. It delivers about 1,361 Watts per square metre to the upper atmosphere. This power is approximately:

- 50% heat (which is also called near-infrared – 700–3,000 nanometres in wavelength [a nanometre is one-billionth of a metre])
- 40% light (all the colours of the rainbow, which look white when blended together – 400–700 nanometres)
- 10% ultraviolet (ionising radiation, which causes skin cancers – 200–400 nanometres).

The Sun's heat radiation (near-infrared) that makes it all the way down to the ground has a wavelength of about 700–3,000 nanometres. As it turns out, hardly any of the gases in the Earth's atmosphere will absorb this kind of radiation. So the Earth's atmosphere is mostly transparent to incoming heat radiation from the Sun.

The result is about 51% of the Sun's heating power reaches the ground as heat (remember, the Sun's radiation *before* it passes through the atmosphere is about 50% heat, 40% visible light, and 10% ultraviolet light).

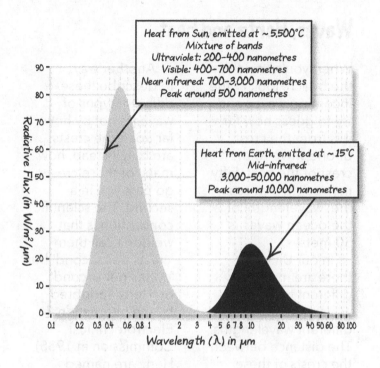

Heat from Sun, emitted at ~ 5,500°C
Mixture of bands
Ultraviolet: 200–400 nanometres
Visible: 400–700 nanometres
Near infrared: 700–3,000 nanometres
Peak around 500 nanometres

Heat from Earth, emitted at ~ 15°C
Mid-infrared:
3,000–50,000 nanometres
Peak around 10,000 nanometres

(Based on graph at climate.be/textbook/chapter2_node3.xml by Y. Kushnir)

It's a very different story with heat radiation that leaves the ground and heads into space. This is called mid-infrared, which has a wavelength of about 3,000–50,000 nanometres.

You guessed the punchline!

This radiation is absorbed by the Greenhouse Gases in our atmosphere. As a result, only about 6% of the Sun's power (as compared to the 51% that hit the ground) will escape our atmosphere and go back into space.

Waves, Hertz and heat

When you're next at the beach, look at the incoming waves and try to guess how far it is from the crest of one wave to the crest of the next. Fifty metres? In that case, the 'wavelength' of the ocean waves is 50 metres.

In our Universe, there are many different types of waves with many different wavelengths. The distance between the crests of these waves can range from thousands of kilometres, to metres, and much smaller: millimetres (thousandths of a metre), or micrometres (millionths of a metre), or nanometres (billionths of a metre), and so on.

Another way to describe these different types of waves is not by how far apart the crests are but, instead, how many of these crests go past you in a second. The scientific convention is that we don't call them 'crests per second' but 'cycles per second', or 'Hertz' (adopted by the International Electrotechnical Commission in 1935). Hertz are named after Heinrich Rudolf Hertz, who was the first person to prove that electromagnetic waves (such as light, heat, microwaves, X-rays, radio waves, ultraviolet, etc.) actually exist.

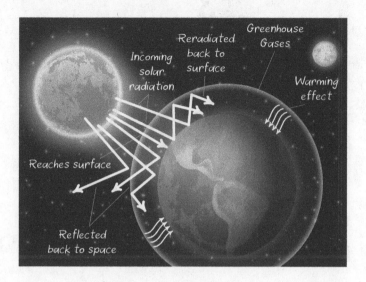

(Based on Greenhouse Effect Diagram by Siberian Art/Shutterstock)

This situation is lovely for us. As I mentioned earlier, prior to the Industrial Revolution, which began around 1750, it gave our atmosphere a nice comfy 14°C temperature.

So, here's the THM (Take Home Message). Greenhouse Gases act like a one-way valve. They let the near-infrared radiation (from the Sun) come freely through the atmosphere, but they prevent most of the mid-infrared radiation (from the ground) from leaving the atmosphere. The heat is trapped, and the temperature rises.

3

BURIED
SUNSHINE

Why do we burn carbon, and where did it come from?

The reason we burned so much carbon over the past 250 years is that it's absolutely loaded with energy. One barrel of oil carries the same energy as 200 kg of coal – which is the same as the amount of energy a person can generate by working for 10 years!

Carbon's been part of the furniture since Earth was born about 4.54 billion years ago. About one-third of a billion years ago, the carbon dioxide levels in the atmosphere were about eight times higher than they are today. Plants sucked most of this carbon dioxide from the atmosphere and used enormous amounts of the Sun's energy (hoorah for photosynthesis!) to convert it into more plants. When those plants died, this carbon was buried in the ground – for a few hundred million years. Then we found this 'buried sunshine', and discovered it was the cheapest source of energy on the planet – even cheaper than horses or slaves, who, after all, have to be fed.

CARBON GETS AROUND

Carbon is the sixth atom in the periodic table. But it wasn't even around when the Universe began with the Big Bang.

The first carbon atoms were made inside the first generation of stars, probably about 280 million years after the Big Bang. They were a by-product of nuclear burning, and formed when three helium nuclei slammed into each other almost simultaneously. After a while, those early stars exploded, throwing carbon atoms into space. Some of those carbon atoms ended up in so-called 'stellar nurseries' – where there's enough 'stuff' (atoms and maybe some small molecules) in so-called 'empty' space for stars to form. Eventually, these atoms were incorporated into the next generation of stars – and into the planets that formed out of the random debris that orbited these fresh stars.

Back in the early Carboniferous period, about 360 million years ago, the carbon dioxide level in the atmosphere was eight times higher than it is today – around 3,000 parts per million (ppm), as compared to the 415 ppm in early 2021. Plants had left the oceans to colonise the land about 450 million years ago, so they evolved to mop up carbon dioxide to use as a building and structural material.

Luckily, photosynthesis (which turns the energy in sunlight into another form of energy that plants can use) had evolved more than 3 billion years earlier. The plants used the energy in sunlight to convert carbon dioxide (low energy, in the atmosphere) into glucose molecules (high energy, inside the plant). Compared to carbon dioxide, glucose (a carbohydrate) carries a huge amount of potential energy. (The word 'carbohydrate' literally means carbon that is 'hydrated' – in other words, carbon [symbol C] is joined to a hydrate [water, which has the symbol H_2O].)

$$6CO_2 + 6H_2O + sunlight \rightarrow$$
$$C_6H_{12}O_6 \text{ (glucose)} + 6O_2 \text{ (oxygen)}$$

Plants then joined these glucose molecules together in long chains to make cellulose, a structural material. These molecules were still carbohydrates.

Back then, the vast Carboniferous forests pumped out so much oxygen, its level in the atmosphere was a staggering 33% (rather than the 21% we enjoy today).

Mega insects!

Unlike us, insects don't have lungs. They get oxygen by diffusion directly into their body. So, a higher oxygen level meant that , during the Carboniferous period, some flying insects could grow to huge sizes - up to 70 cm across!

Meganeura: a genus of extinct insects related to present-day dragonflies and damselflies.
(Dodoni/Wikimedia Commons)

Etching depicting some of the most significant plants of the Carboniferous.
(Bibliographisches Institut – Meyers Konversationslexikon)

More important to us is this: how were those carbon atoms recycled into fossil fuels such as coal (3,510 billion tonnes still accessible), oil (230 billion tonnes) and methane gas (140 billion tonnes)? The answer is biology, geology and chemistry.

At the beginning of the Carboniferous period,

bacteria and fungi that could digest wood had not yet evolved. So dead trees didn't fully rot and, over millions of years, vast forests of dead trees ended up being buried underground virtually intact. At least, that's one theory. Other theories say that trees grew and fell over in wetlands, or that atmospheric temperatures (or moving tectonic plates) were other significant factors in forming fossil fuels.

Now, add hundreds of millions of years to the mix. Underground, the carbohydrates in the cellulose were gradually converted into hydrocarbons, still loaded with huge amounts of potential energy. (Hydrocarbons are made only from hydrogen and carbon atoms – the oxygen atoms have gone.) We're not 100% sure of all the details, but we are confident that lower temperatures (35–80°C) underground turned plants into lower-quality coal (like brown coal), while higher temperatures (180–245°C) gave us high-quality anthracite coal.

So carbon atoms were taken from the air into plants via biology, put underground via geology, and finally incorporated into coal via chemistry.

Biology, geology and chemistry also gave us oil (called petroleum in the USA). But oil came to us from different starting materials – fossilised organic materials, such as algae and zooplankton.

On planet Earth, there are now about 60 million billion tonnes of carbon in the form of coal, oil, gas, etc. in the 'lithosphere', the rocky, gloopy layer that extends down from Earth's surface to about 80 km deep. Over hundreds of millions of years, a small fraction of these carbon atoms were recycled into the oceans (38,400 billion tonnes) via geology, through hydrothermal vents on the ocean floor. Once these carbon atoms were in the 'biosphere', living creatures could start using them as constituents of food.

ENERGY AT BARGAIN-BASEMENT PRICES

Why did we ever start burning these carbon-rich fossil fuels? Well, the reason is straightforward and simple – fossil fuels are absolutely loaded with ridiculously cheap energy.

One 159-litre barrel of oil costs between A$50 and A$100. This barrel (or 200 kg of coal) carries as much energy as the average person can generate over more than ten years of physical labour (40 hours per week, 48 weeks per year). The current average annual salary for a labourer in Australia is around A$50,000, so ten years' labour costs A$500,000 – or just A$50 if you burn oil instead.

What a bargain!

But suppose we don't spread out the burning of that barrel of oil over ten years, but do it much more rapidly, in 20–30 seconds. In that case, which is the equivalent of burning two or three barrels of oil each minute, we can generate enough power to fly a 600-tonne aircraft carrying 500 passengers at about 900 km/h. If we keep doing this for about 21.5 hours (or about 1,300 minutes), we have flown from Sydney to London – and burned the energy equivalent of more than 3,000 barrels of oil.

What amazing power!

So what about powering a car? Well, an SUV's tank will take 70 litres of petrol (which carries the energy equivalent of one person working hard for five years). Petrol comes from oil, which originally came from algae and the like. But let me give a more relatable example of plants on land. To generate the energy present in that 70 litres of petrol would take 1,640 tonnes of vegetation, which grew on about 1.13 km^2 of land, followed by several hundred million years of geology and chemistry.

You can see why fossil fuels deserve the nickname 'buried sunshine'.

To extract this enormous amount of energy, all we had to do was run the chemical reaction backwards, by the ludicrously simple method of burning.

$$C + O_2 \rightarrow CO_2 + \text{(enormous amounts of energy)}$$

Burning wood was pretty easy – we just had to follow the example of lightning, which has been starting forest fires since there were forests. But I still wonder who the first person was to realise that some black rocks (coal) could burn.

Oil saved the whales

As an aside, crude oil helped to save the whales from almost certain extinction. Virtually every use of whale oil – including heating and lubrication – could be covered by a by-product of crude oil.

But sperm whale oil still had one very specific use – lubricating the sliding blocks of subcritical uranium-235 in an atom bomb. The claim was that sperm whale oil was the only oil that would not dry out after several years of inactivity. I've also read the specification sheet for the World War II 6x6 Studebaker truck, which specifies using sperm whale oil for the moving parts of the booster in the power brakes.

In Australia, whaling finished in about 1978. But when I visited an old whaling station at Albany in Western Australia in 1989, I noticed the tourist shop had a small bottle of red sperm whale oil for sale. I bought it. I gave some of my precious sperm whale oil to a friend, who was restoring a Studebaker truck. But don't worry – I don't have any old atom bombs that need oiling.

We humans started recovering some of this concentrated Sun's energy on a big scale during the Industrial Revolution, which began in about 1750.

The essence of the Industrial Revolution was making stuff on a huge scale, but, for the first time in history, using labour from machines instead of animals. These machines were powered by steam, which was generated by burning carbon to heat water. While wood could burn, you needed lots of it and, anyway, wood was also essential for construction (for ships, buildings, etc.). So coal, with about twice as much energy per kilogram as wood, and no other clear use, was the obvious fuel. Until it wasn't ...

4

400,000 ATOM BOMBS AWAY!

That's how much extra heat we're trapping a day — no wonder it's getting hot!

Thanks to all those fossil fuels we're burning, the carbon dioxide level in the atmosphere is rising at an alarming rate. We know — we've been measuring it every day for more than 50 years. And the extra carbon dioxide that we've added is trapping a lot of heat — 400,000 atom bombs worth of heat a day!

In the early days, the major emitters of Greenhouse Gases were first Europe, followed by the USA after a century or so. Counting since the Industrial Revolution, Europe still holds the record for the greatest quantity of emissions dumped into the atmosphere. But, as of 2019, the biggest single current emitter is China.

UP, UP AND AWAY

In 1860, about a hundred years after the Industrial Revolution began, Irish physicist John Tyndall was measuring how well carbon dioxide absorbed heat compared to other gases. Tyndall's original equipment is still on show at the Royal Institution in London. His set-up was fragile and labour-intensive. His apparatus was not suitable for continuous use by non-scientists, nor was it weatherproof.

Over the next century, scientists developed better instruments.

Using Tyndall's method, but with better instruments, we have directly measured carbon dioxide levels in the atmosphere all day, every day, since 1958. This was when Charles David Keeling started his measurements of carbon dioxide concentration at Mauna Loa Observatory, in Hawaii, some 3.4 km above sea level and far from any industry (which would emit carbon dioxide), vegetation (which could absorb carbon dioxide in spring), large land masses, etc. It was measuring global, rather than local, levels.

In pre-industrial times (before we started burning carbon that had been underground for hundreds of millions of years), the carbon dioxide level was about 280 ppm. We know this by measuring the carbon dioxide levels in bubbles of pre-industrial air trapped inside ice cores, and so on.

Keeling's equipment measured the carbon dioxide levels in the atmosphere rising from 313 ppm in 1958. It passed 400 ppm on 9 May 2013. As of early 2021, the carbon dioxide level measured around 415 ppm. Currently, it's rising at over 2 ppm each year.

When these measurements are plotted on a graph, they show the famous Keeling Curve (you can see this graph on page 18). Keeling personally supervised this monitoring program until he died in 2005. His son, Ralph Keeling, then took over the program.

There are a few 'features' to the Keeling Curve.

First, it's one of the most important scientific achievements of the 20th century.

Second, the general trend is relentlessly upward.

Third, within each calendar year, there's an extra oscillation, which gently curves up and down by about plus, then minus, 2 ppm. This is caused by the plants across the world taking in more (or less) carbon dioxide as the seasons roll through the year and is dominated by the Northern Hemisphere's land mass.

Fourth, modern measuring equipment is so sensitive it can even measure a tiny bump upwards in carbon dioxide at night (when plants respire and emit carbon dioxide) and another slight bump downwards in the day (when plants use carbon dioxide as a raw material).

Before Keeling's measuring equipment on Mauna Loa, there were only scattered and non-continuous direct recording of carbon dioxide levels on Earth.

Today, Greenhouse Gas levels are measured at about 100 measuring sites around the world. This network is called the Global Greenhouse Gas Reference Network, and measures carbon dioxide, methane and nitrous oxide, as well as carbon monoxide to estimate air pollution.

GLOBAL CO_2 EMISSIONS OVER TIME SINCE 1750

For the first century after 1750, annual carbon dioxide emissions from human activity were less than 1 billion tonnes per year. Back then, coal was the major fossil fuel, and we can accurately estimate carbon dioxide emissions based on how much coal was mined around the world.

After this slow start, however, carbon dioxide emissions reached 2 billion tonnes per year by 1900, and 5 billion tonnes per year by around the 1920s.

In about the 1950s, there was an abrupt increase; and, by around 1970, carbon dioxide emissions reached 12 billion tonnes per year. By the year 2000, emissions were at about 25 billion tonnes per year. By 2010, more than 30 billion tonnes of carbon dioxide were being emitted each year.

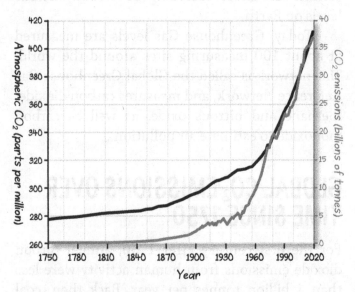

CO_2 in the atmosphere and annual emissions (1750–2019)
(Data via National Oceanic Atmospheric Administration)

Obviously, as we burned more carbon, carbon dioxide levels in the atmosphere rose (although plants, oceans and so on soaked up some of the carbon dioxide). So here we are today, with some 1,600 billion tonnes of carbon dioxide emitted since 1750.

(By the way, emissions from volcanoes are consistently about 1% or less of our current human-related emissions. It is a 'mistruth' to say it's the other way around.)

400,000 HIROSHIMA ATOM BOMBS EVERY DAY, ONE TSAR BOMBA EVERY 12 MINUTES, ONE DINOSAUR-BUSTER ASTEROID EVERY 20 YEARS!

The amount of extra heat from the Sun trapped by the current level of Greenhouse Gases is about 400,000 Hiroshima atom bombs each day. That's an incredible, but dreadful, number. However, this energy is spread over the 510 million km^2 of the entire planet's surface, not just concentrated into a single square kilometre.

A single Hiroshima atom bomb has the energy of about 16,000 tonnes of the explosive TNT (trinitrotoluene). We're trapping this huge amount of energy into the atmosphere – four or five atom bombs each second, for 86,400 seconds each day, 365 days a year ... you get the picture.

Let's take it up a level.

The most powerful nuclear weapon ever exploded was the 50-megatonne Russian Hydrogen Bomb – the Tsar Bomba. In a fraction of a second, it released the energy of 50 million tonnes of TNT. The flash of light was visible 1,000 km away, and the mushroom cloud it produced rose 65 km high – about six times higher than the typical cruising altitude of a passenger jet. The column of the mushroom cloud was 10 km across, the mushroom cap was 90 km wide, and it was visible from 800 km away. If Tsar Bomba exploded in Melbourne, you could see the top of the mushroom cloud from Sydney, some 700 km away. At ground level, the zone of total and absolute destruction was 70 km across.

In our atmosphere, Greenhouse Gases capture one Tsar Bomba of heat energy every 11–12 minutes. That is huge.

Let's now take this to the max.

Around 66 million years ago, a 10-km asteroid helped wipe out the dinosaurs when it smashed into the Earth. The impact site is at what is now the northeast corner of the Yucatán Peninsula, at the southern end of the Gulf of Mexico. The crater is about 20 km deep (over twice the height of Mount Everest) and 150 km across.

In our atmosphere, carbon dioxide behaves like a one-way valve. It lets most of the short-wavelength infrared heat from the Sun in through the atmosphere until it hits, and heats, the ground and oceans. The Earth's surface then radiates some of this heat (now at a longer wavelength) upwards, back towards space. But carbon dioxide (and other Greenhouse Gases) catch some of this outgoing heat.
(Based on illustration by A Loose Necktie/Wikimedia Commons)

The asteroid's impact set off massive tsunamis hundreds of metres high. The blast wave killed all unprotected life for several hundred kilometres around. Earthquakes and volcanoes were set off around the entire Earth.

Several trillion tonnes of debris (more than the mass of the asteroid) were blasted up and outwards. Some went into space and were gone forever, while the rest landed back on the ground. Today, the debris layer is up to 80 metres thick within 500 km of the impact site, and 2–5 mm thick on the other side of the planet.

Think of how much energy that would have taken to do all that. Now consider that, over the past 20 years, from 2000 until 2020, Greenhouse Gases have captured one Dinosaur-Buster asteroid's worth of heat energy.

That's a lot of energy to put into a planet-wide biosystem – so it's no surprise there are massive consequences to Global Warming. But before we get to those, we need to know where the emissions are coming from.

WHO'S EMITTING WHAT, NOW?

The source of carbon dioxide emissions has changed over time. Back in 1750, when the Industrial Revolution began, first Britain and then Europe had the highest emissions in the world. They later passed the baton to the USA. It took longer again to get to Asia.

By 2018, the largest carbon dioxide emitter was China, with 28% of the global total. It was followed by the USA (15%), Europe's 28 member states (9%), India (7%) and Russia (5%). Between them, these five 'regions' accounted for about 64% of global carbon emissions.

Shipping-at-sea (marine bunker fuels) and aviation fuel accounted for about 5% of global emissions.

Australia came in with 1.1% of global emissions – even though we have only 0.3% of the global population. After the Arabian countries, we are one of the top per-capita Greenhouse Gas emitters on the planet. Per person, our carbon footprint is about 11 times higher than India's, one-and-a-quarter times higher than the USA's, and three times higher than China's. And that's just what we emit domestically. When you include our fossil fuel exports, Australia's global emissions footprint is about 3.6% of the world total.

WHAT ABOUT CUMULATIVELY?

Cumulative global emissions since 1751 by region

In total, we humans have emitted some 1,600 billion tonnes of carbon dioxide into the atmosphere since the Industrial Revolution. Not all of it stayed there – some went into plants, the soil, the oceans, etc. This is a listing of where it all came from – from 1750 until 2019.

In terms of combined cumulative emissions since 1751, 'Europe' (which, in addition to the 28 member states of the European Union [EU], also includes Russia, Ukraine, Norway, Turkey, and a few others) leads, with 33%. This makes perfect sense. These countries got into burning fossil fuels before everybody else, and they have big populations.

Equal second, with 29%, are 'North America' (which includes Canada and Mexico) and 'China and Other Asia' (which also includes Iran and Lebanon). Again, this sounds perfectly reasonable. North America started a bit later than Europe, but it overtook Europe in terms of consumption and wealth. China (and 'Other Asia') started serious burning of fossil fuels later again – but their population is much greater. Between them, China and India have about one-third of the population of the entire world (about 1.4 billion people each).

All the nations south of the Equator (Africa, South America and Oceania) account for a bit less than 7%. This is in part because the majority of the world's people live in the Northern Hemisphere. As well, Africa and South America have always lagged behind the rest of the world in terms of wealth and consumption. Still, I am a little surprised that only 7% of the world's total cumulative carbon dioxide emissions came from the southern half of the world ...

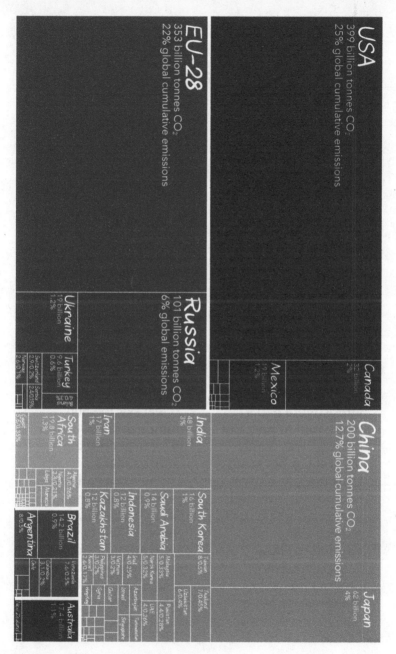

Cumulative emissions by region.

(Creative Commons. Data calculated by Our World in Data based on data from the Global Carbon Project [GCP] and Carbon Dioxide Analysis Center [CDIAC])

Where does our 'warmth' come from?

Way back in 1824, Joseph Fourier grappled with the then-unsolved problem of where the Earth's 'heat' comes from. After all, the Earth (+14°C) is so much warmer than the Moon (-23°C). Fourier identified potential heat sources such as the Sun, the core of our planet, and the rest of the Universe – and he also cleverly realised that the Earth's atmosphere had a major part to play. But his work was purely theoretical, because, back then, scientists had not yet invented instruments to actually measure what he needed to know.

First, it turns out that the rest of the Universe has a temperature of about -270°C (that's *minus* 270°C) – so it's sucking heat from us, not giving it to us.

Second, the 'waste heat' from the decay of radioactive elements in the Earth's core adds another 0.1 Watt/m² (that's about 30 times smaller than the heat caused by Greenhouse Gases).

However, when the Sun is directly overhead and there's a clear sky, the Sun dumps about 500 Watt/m² of heat onto the Earth's surface.

The extra carbon dioxide (and other Greenhouse Gases) we've added to the atmosphere since the Industrial Revolution create an extra 3.1 Watt/m².

The 'waste heat' from car engines, air conditioning, etc. adds another 0.028 Watt/m². That's about 100 times less than the heat caused by Greenhouse Gases.

So, Greenhouse Gases cause the vast majority of our current episode of Global Warming.

CARBON EMISSIONS BY SECTOR

Another useful way to analyse Greenhouse Gas emissions is to break it down by 'sector', such as transport, agriculture or manufacturing. This way we can see which human activities are generating what emissions, so we can start to think about how we can change the way we do things to reduce them.

There are lots of different ways of breaking up the sectors, but here's a pretty straightforward one.

This information is incredibly important to get your head around, because once we know the individual sources of Greenhouse Gas emissions, we know where to apply the fix. (Yes, knowledge is power.) In chapter 8, we'll look at exactly that – how to fix the way we do things in some of the various sectors, so we can cut emissions fast.

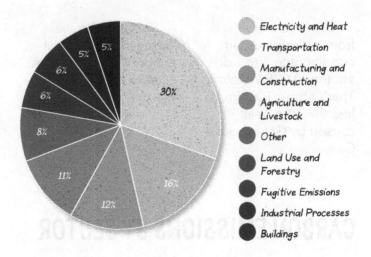

Greenhouse Gas emissions by sector
(Data by Climate Watch)

Legend:
- Electricity and Heat
- Transportation
- Manufacturing and Construction
- Agriculture and Livestock
- Other
- Land Use and Forestry
- Fugitive Emissions
- Industrial Processes
- Buildings

Electricity and heat

Emissions from power generation account for around 30% of global Greenhouse Gas emissions. That's a big chunk!

Transportation

Sixteen per cent of global Greenhouse Gas emissions come from burning fossil fuels to power transport, including road transport (which accounts for around 75% of overall transport emissions), air, shipping and rail. These don't include emissions from the manufacturing of motor vehicles or other transport, which are included in Electricity and Heating above.

Manufacturing and construction

Emissions caused by construction, and the manufacture of things such as cars, textiles, wood products, etc., currently account for around 12% of Greenhouse Gas emissions.

Agriculture and livestock

Growing and burning crops, as well as rearing farm animals for food or other uses, causes around 12% of emissions. Cows' burps, farts and poos account for around half of that, at a 'stinking' 6% of all global emissions!

Other

Eight per cent of global emissions are caused by the burning of other fuels for various purposes.

Land use and forestry

Land clearing and the timber industry cause around 6% of emissions, taking into account that replanting trees is treated as negative emissions. However, while a tree can be chopped down in minutes, it can take 40 years to regrow.

Fugitive emissions

These are caused by the often-accidental leakage of methane to the atmosphere during coal mining, fracking, and oil and gas extraction and transportation, but they still add up to 6% of emissions.

Industrial processes

Industrial processes account for about 5% of emissions. This is a slightly 'fuzzy' sector, but usually means 'emissions from the smokestacks of industrial plants'. There is some overlap with other sectors; it typically excludes burning fuels for energy, but includes the emissions from chemical reactions used in production (e.g. of paper, chemicals, food, etc.).

Buildings

'Building emissions' are 'embodied' carbon emissions not accounted for elsewhere, or carbon associated with materials, construction and maintenance processes throughout the entire building lifecycle. It's estimated that these emissions associated with buildings account for around 5% of global emissions.

Carbon emissions – all over the place

Coal was our first major fossil fuel.

We began to use oil as a fossil fuel in significant quantities in about 1920. Because it was liquid, it was easier to transport. Both coal and oil generate Greenhouse Gases only when they're burnt.

But **methane**, also known as natural gas (or CH_4), is a Greenhouse Gas in its natural state – and it is much more damaging than carbon dioxide. When it's burned to generate heat, the by-product is carbon dioxide. (So, surprisingly, burning methane to make energy causes less greenhouse warming than simply releasing it into the atmosphere.)

$$2CH_4 + 4O_2 \rightarrow 4H_2O + 2CO_2 + \textbf{(enormous amounts of energy)}$$

Methane was originally seen as an unnecessary and useless by-product, and was simply vented into the atmosphere. However, we began to use it as a fuel in significant quantities in about 1950.

Then there's 'cement'. Of course, it's not a fuel, but I am including it because its production creates a whopping 8% of carbon emissions! Cement is a 'glue' or 'binder' that has two useful properties. First, it can set and harden, and, second, it can stick to other materials to join them together. If you mix cement with coarse aggregates (such as sand and gravel) you get concrete, but when mixed with fine aggregates you get mortar (which is used to join bricks together).

Cement has been made for thousands of years. The basic way to make cement is to apply massive heat to limestone to make quicklime (also called lime). The lime is then combined with other chemicals (silicates, aluminium oxide, etc.) to make hard 'rocks' called clinker. This clinker is then ground to make the fine powder called 'cement'.

$$CaCO_3 + \text{(enormous amounts of heat)} \rightarrow CaO + CO_2$$

This reaction needs temperatures of around 1,450°C. Generating this heat accounts for about half of the Greenhouse Gas emissions of cement, while the carbon dioxide given off accounts for the other half. About 900 kg of carbon dioxide are emitted for each 1,000 kg of Portland cement made. If the cement industry were a country, it would be the third-largest producer of Greenhouse Gases – after China and the USA.

There already are alternatives to standard cement that have much lower carbon footprints. One very promising alternative is 'geopolymer'.

Steel is another major producer of greenhouse emissions (around 8%) – but like cement, it can easily be brought to virtually zero.

Australia is the world's major producer of iron ore (over 800 million tonnes in 2015, more than double that of the output of Brazil). However, we make hardly any steel with this iron ore.

The major production of Greenhouse Gas emissions happens in the chemical reaction where iron oxide (the raw dirt that is mined so much in Western Australia) is chemically turned into iron. The current chemistry is that carbon (usually coke, that is made from coal) is added to iron oxide, and heat is applied to start the reaction. (There are several types of iron ore, such as FeO, Fe_3O_4, Fe_2O_3, $FeCO_3$, etc. – but I'll keep it simple with FeO.)

$$C + 2FeO + \text{(small amounts of heat)} \rightarrow 2Fe + CO_2 + \text{(huge amounts of heat)}$$

Carbon is absolutely brilliant here. First, carbon can drag the oxygen atom away from the iron – if the temperature is high enough to set the reaction going. Second, this reaction (once you get it started) generates huge amounts of heat. The only downside is that it releases carbon dioxide, which is a Greenhouse Gas.

However, hydrogen can do exactly the same job, with the same benefits – remove the oxygen atom from the iron atom AND generate lots of heat. And instead of carbon dioxide, it releases water as a by-product.

$$H_2 + FeO + \text{(small amounts of heat)} \rightarrow Fe + H_2O + \text{(huge amounts of heat)}$$

A trial run in April 2020 at a commercial steel mill in Sweden produced steel using hydrogen (instead of carbon) and with no reduction in quality.

This chapter might seem like just a bunch of boring old numbers. But if you can't measure something, you don't know what you're dealing with. And if you don't know the past, you can't plan for the future.

That's why this history is so important. We can avoid making the mistakes of the past.

When I first saw emissions broken down by sector, I realised that solving Climate Change was achievable. Even better, it was relatively easy to do. All we have to do is focus on each sector and work out how to reduce emissions in it. We definitely can reverse Climate Change with the science/ technology we have today, or, indeed, that we had back in 1990. But, while we do not need amazing new scientific breakthroughs to have a happy outcome, these breakthroughs will happen (that's science for you!) and they will certainly help.

That knowledge gave me incredible hope and joy (and anyhow, I don't think any list of numbers is boring).

5

SO HOT IT HURTS THE PLANET

How rising temperatures affect the planet and its environment

The effects of Global Warming on our planet vary enormously across the globe – more at the Poles, and less at the Equator. There are a few reasons for this – including physics and geography.

In fact, Earth has even been tipped off its axis by Global Warming – only a little bit, but still ...

Here's a little essential background info first, and then let's dive into a few examples of how Climate Change is impacting our planet.

First, physics. About 71% of our planet's surface is water – not land. When it comes to absorbing heat, water behaves very differently from land. Physicists tell us that water has a 'high specific heat capacity'. This means that when we dump a lot of heat into water, it will respond by increasing its temperature by only a little bit.

Different substances soak up heat differently. For example, let's say we dump the same amount of heat into the same masses of water, air, dry soil and iron. If the temperature increased by 1°C in the water, it would increase by 4.2°C in air, 5.2°C in dry soil and 8.4°C in iron.

So water 'soaks' up heat about four to five times better than air and soil.

Second, geography. The total surface area of our planet is enormous – about 510 million km^2 of land and ocean. The geography varies from place to place – think deep oceans versus shallow lakes, deserts versus dense forests, mountains versus valleys, canyons versus open plains, and polar versus equatorial regions, etc. After all, the geography of a location has a big impact on its local climate.

More importantly, the Northern Hemisphere has more of the land area (two-thirds) than the

Southern Hemisphere (one-third). (It was the other way around about 300 million years ago, due to the continents migrating on the tectonic plates.) So Climate Change affects each hemisphere a little differently.

Taking the above into account, let's look at four effects of Climate Change on climate bands, the oceans, the land and, finally, Earth's axis – because our entire 12,700-km-in-diameter planet has been tipped off its axis!

MIGRATING CLIMATE BANDS AND THE POLES

Climatologists are very confident that the Earth's air temperature (measured at the surface) has risen by about 1°C, averaged across the globe, since the beginning of the Industrial Revolution in about 1750.

But that's just the overall general trend. Let's look at it more closely.

There are different bands of climate around the Earth, such as the tropics, subtropics, temperate regions, and so on. These climate bands are all moving away from the Equator towards the Poles. They're doing this at an average speed of about 5 km per year, or 50 km per decade, or 500 km per century. As a result, temperature and rainfall patterns are shifting. This means that growing conditions will change for our farmers.

For example, farmers grow crops that suit the local soil, rainfall, sunlight, temperature range, etc. If the local climate changes, farmers have only a few options. They can sell up and shift to another area, or they can stay and use genetically modified crops to suit the new conditions, or grow different crops entirely. But these are major changes, and they don't happen overnight. Major transitions can be complex, tricky and sometimes don't turn out the way we want.

Another issue related to the impact of Global Warming is that the temperature rises much more in the polar regions than it does near the Equator. As well, there's a major difference between the northern and southern polar regions when it comes to the temperature increase.

To understand this pattern, you need to know that the Antarctic is the 'opposite' of the Artic. The Arctic has a small amount of sea and pack ice, floating on water, closely surrounded by continents (North America, Europe and Asia). The average thickness of the ice was about 3.5 metres back around 1980, but it's now much thinner at 1.3 metres.

In contrast, the Antarctic is a huge continent, roughly twice the size of Australia. It's far from neighbouring continents. It has a huge amount of ice with an average thickness of 2,160 metres but up to 4,776 metres. This ice is sitting on land, not floating on water.

So up around the North Pole, there's a little ice surrounded by lots of land and a relatively small amount of water. But around the South Pole, there's heaps of ice and not much exposed land surrounded by huge amounts of water. Remember, water can absorb lots of heat before it changes temperature very much.

In the Arctic region, air temperatures are increasing at twice the rate of the rest of the world. By the way, white ice reflects the incoming heat while the darker water will soak up the heat. The area of the white reflective ice has been dropping by about 13% each decade since 1980. This is leading to a positive feedback loop (see pages 84–85): ice melts → more dark water → water absorbs more heat → melts more ice → more dark water, etc.

At this stage, we have not yet reached the irreversible positive feedback loop. We can still save the Arctic ice.

At the other end of our planet, the Antarctic is surrounded by the vast Southern Ocean. You'd expect the water (with its high heat capacity) to buffer the heat from Climate Change, so that the temperatures there would be more stable. But it hasn't!

What's a Positive Feedback Loop?

A 'feedback loop' is a process that happens within a 'system', where the output affects the input, which affects the output, which affects the input – and so on, until the system settles into a new state or becomes unstable. Feedback loops are everywhere – in the burning of stars, stampeding cattle, panic-buying of toilet paper, the high-pitched squealing in an audio system when the microphone is too close to the loudspeaker, thermal runaway in electronic circuits that can't get rid of the heat, etc.

A feedback loop can be negative or positive.

An example of a negative feedback loop is when it gets hot, and you start to raise your body temperature over your regular 37°C (the input). After a while, you begin to sweat (the output). This sweat evaporates, which takes heat away from your body and you begin to cool down. This lower temperature is now an input, but because it's low, it does not make you sweat.

So in this negative feedback loop, the output (the sweat) acted to reduce the input (your body temperature).

But in a positive feedback loop, the input acts to make the output bigger, by making the input bigger.

A stampede is an example of a positive feedback loop. In a large herd of cattle, one becomes nervous or alarmed (the first input). This then alarms two or three neighbouring cattle (the first output). The alarmed state of these several cattle (this is now the second input) excites ten more head of cattle (the second output, which is a lot bigger than the first output). And so the state of excitement or alarm spreads through the herd, successively affecting a larger number of cattle. The output (the number of excited cattle) increases until all 1,000 head of cattle are stampeding off into the night – a sad case of positive feedback.

In the case of permafrost, a little bit of global warming heats up the permafrost until it releases its Greenhouse Gases. These then cause more Global Warming, which causes the permafrost to release more Greenhouse Gases, which causes more Global Warming – you get the idea. It's a positive feedback loop that finishes only when all the Greenhouse Gases have been released from the permafrost *or* we humans take action to interrupt this positive feedback loop.

Climate Change has given us some nasty surprises. Changes in water currents have happened in the Southern Ocean.

In January 2020, the Australian Casey Research Station in Antarctica recorded its highest-ever air temperatures since the base was established in 1964, with a daytime maximum of 9.2°C and a night-time minimum of 2.5°C.

In February 2020, the Argentinian Research Station in Antarctica, Marambio, also reported its highest-ever air temperature of 20.75°C.

Normally, almost all of Antarctica is below zero, all day, every day. Not anymore. Temperatures above zero can easily melt land ice. The ice-cream cone of Antarctica is dripping away.

OCEANS

'Luckily' for us, oceans are where all the heat's going

On one hand, the scientists tell us that the lower atmosphere has warmed by roughly 1°C since pre-industrial times. But you would expect that adding 400,000 Hiroshima bombs of greenhouse heat each and every day to the atmosphere would have warmed it up by *more* than just 1°C.

In fact, this amount of energy 'should' have warmed up the atmosphere by 35°C to 40°C.

'Luckily' for us, the heat does not get stuck in the lower atmosphere. About 93% of it goes into the oceans – more in the top layers of the oceans and less in the lower layers. (But, given enough time, the ocean currents will carry this heat into the ocean's deeper layers.)

So, in the short term, we are not all dead because the oceans soak up most of the heat from Global Warming.

Think back to 18 December 2019, for example. On that day, the average peak temperature across Australia (from Cape York to Tasmania) was 41.9°C. Imagine if it had been 80°C!

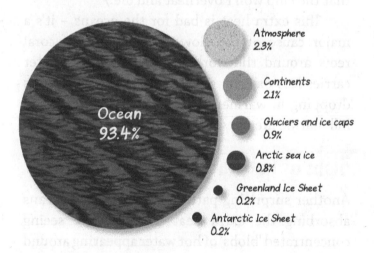

Ocean
93.4%

Atmosphere
2.3%

Continents
2.1%

Glaciers and ice caps
0.9%

Arctic sea ice
0.8%

Greenland Ice Sheet
0.2%

Antarctic Ice Sheet
0.2%

The amount of global warming going into various components of our climate system, from 1993 to 2003.
(Based on graph by John Cook/Skeptical Science, data from IPCC AR4 5.2.2.3.)

What about 4 January 2020? On that day, the city of Penrith, in Sydney, was the hottest place on Earth – 48.9°C. Imagine if it had been 85°C!

But there's a BIG downside

However, the oceans can't be our heat 'sink' indefinitely. (In science and engineering, the word 'sink' means a place you can dump stuff – pollutants, heat, etc. – so it effectively 'vanishes' from your local system. Computers often have a 'heat sink' mounted on a processor chip to help carry away the heat to outside the computer, so that the chip won't overheat and die.)

This extra heat is bad for the oceans – it's a major cause of the slow-motion dying of coral reefs around the world. As well, warmer water carries less oxygen, so we're seeing oxygen levels dropping in warmer waters – which by itself can kill marine life.

Blobs of hot water

Another surprising pattern caused by the oceans absorbing the extra heat is that we're seeing concentrated 'blobs' of hot water appearing around the world's oceans. In 2014, a blob of hot water appeared off the USA's Pacific Coast, some 1°C to 4°C hotter than normal. It lasted for two years, leading to the unprecedented deaths of salmon and sea lions.

'Blobs of hot water' actually have an official name – Marine Heat Waves (MHWs).

Ships first observed MHWs in the 1920s, but from about 1980, we could measure them much more accurately via satellites. Looking at averages, we've found that back then more than 90% of MHWs observed lasted fewer than 14 days and covered an area less than 200,000 km^2. But the 300 largest MHWs since 1980 have averaged 40 days' duration and covered 1.5 million km^2.

This really surprised me. Somehow, I'd thought that the oceans would be 'well-mixed', with no sudden or major shifts in temperature, or localised blobs of hot water. But I was wrong.

These blobs of hot water have been popping up frequently and spreading across the globe to places as far apart as the Northwest Pacific Ocean, the Indonesian–Australian Basin and the Southern Ocean.

Before the Industrial Revolution, MHWs happened only very rarely – every 100–1,000 years or so for the North Atlantic, and every 10,000 years or so for the rest of the world.

Unfortunately, 93% of that 400,000 Hiroshima bombs' worth of heat goes into the oceans each day. A direct result is there have been more than 30,000 MHWs in just the past four decades.

In the 1980s, there were only 27 really big MHWs recorded – with an average over-temperature of 4.8°C, and an average duration of 32 days.

But in the 2010s, there were more (172, not 27), they were hotter (5.5°C above average, not 4.8°C), and they lasted longer (48 days, not 32 days).

So, as the heat energy has piled up in the oceans, MHWs have become more frequent, hotter and longer.

The effects of MHWs are devastating. Sea birds, sea mammals and fish die. Ocean productivity is reduced, so fishing vessels get smaller catches. Harmful algal blooms and substantial coral bleaching events occur. Sea grass and kelp are wiped out, marine biodiversity drops, tropical fish communities move towards the Poles, fisheries close, there are substantial drops in sea ice, and much more.

All these events occurred in 2020, with an atmosphere just 1°C warmer than pre-industrial times.

Things get much worse if the world is 3°C warmer. Projections show that the Northeast Pacific Ocean, Southwest Atlantic Ocean and Indonesian–Australian Basin would be 'essentially in a continuous, extreme heatwave state', according to a 2020 *Science* paper. Eventually, the remaining marine life would evolve to adapt to these higher temperatures. But that would take centuries, or millennia – and these changes are happening over a much shorter time scale of decades.

Unfortunately, we've already killed so much life in our oceans that the remaining marine biomass is about equal to the mass of all of our shipping (about 2 billion tonnes fully loaded). We don't want a 3°C-warmer world, because then that marine biomass will plummet even further.

Coral bleaching

MHWs (and other factors such as oxygen deprivation, local land pollution run-off, etc.) have led to *mass* coral bleaching. These are bleaching events that affect entire reef systems, not just a few isolated coral reefs.

Before the early 1960s, marine scientists had never recorded one mass coral bleaching in the whole world. In Australia, between 2015 and 2020, we had three mass coral bleachings of the Great Barrier Reef. The most recent, in March 2020, was the worst ever recorded – but it was under-reported because of the COVID-19 crisis.

We have to act soon to save the Great Barrier Reef. Why? Food and money.

On the food security issue, coral reefs make up about 0.1% of the surface area of the world's oceans, but provide 25% of all the ocean fish species.

When it comes to money, the economic advisory company Deloitte estimated that the 'total asset value' of the Great Barrier Reef in 2017 was A\$56 billion (including contributing A\$6.4 billion to the national economy in 2015–16 and supporting 64,000 jobs). (And yet, according to a 2019 International Monetary Fund working paper, Australian state and federal governments spend A\$47 billion each year 'to prop up fossil fuel extraction and energy production'. Does that mean we are paying fossil fuel companies to kill off the money-making asset of the Great Barrier Reef?)

The next level up from a *mass* coral bleaching is a *global* coral bleaching event. This is a set of mass coral bleaching events across the three tropical ocean basins – the Pacific, Atlantic and Indian oceans.

Before 1997, there were none. The first global coral bleaching event took place during 1997 and 1998. In that event, 16% of all global reefs died. The second occurred in 2010. The third began in 2016. The long-term effects (e.g., permanent coral dieback) will be painfully apparent over the next decades.

Coral reefs around the world are in a very fragile position.

Currents shift

In some ocean areas, the warming of the waters has changed the traditional ways that the currents used to flow. Some of these currents carried plankton and phytoplankton that fish would eat. But with their food supply no longer there, some fish stocks are declining.

So ocean currents can shift, just as temperature bands on land can shift.

Ice melts …

When the ice that is sitting on land melts, the resulting water eventually runs into the oceans. This is a major contributor to ocean-level rise. Another contributing factor to rising ocean levels is the rise in ocean temperatures. Hot water expands and takes up more volume than cold water.

Sea levels are currently rising worldwide at an average of about 3 mm per year, and this rate is increasing. But on top of the predicted average sea-level rises, there have been some deeply disturbing surprises.

In mid-2019, during the Northern Hemisphere summer, so much ice melted in Greenland that, in just two months, the global ocean level jumped 2.2 mm. In that short summer, the amount of ice that melted in Greenland alone was roughly equal to what recently used to melt across the world in a whole year – nearly 600 billion tonnes.

Those 2019 new summer losses are more than double Greenland's 2003–2016 yearly average.

Elsewhere in the world, low-lying parts of Florida near the coast now flood regularly with each king tide, something that didn't happen two decades ago.

It seems that the total volume of land ice that melted in 2020 was the highest ever – 750 billion tonnes over the whole globe.

More acidic

The oceans have become more acidic by pH 0.1 (thanks to extra carbon dioxide dissolving in sea water). Because of this increased acidity, some sea creatures can no longer make their shells (read more about this on pages 113–120).

Upper atmosphere cooling!

Yup, the upper atmosphere (80 km above sea level) is definitely cooling.

This makes perfect sense. Here's why ...

A certain amount of heat is coming up from the ground.

Greenhouse Gases trap some of that heat before it gets into the upper atmosphere.

Before we dumped those extra Greenhouse Gases into the atmosphere, more of that heat passed through the upper atmosphere on its way into space. On its way out, a tiny amount of that heat would warm the upper atmosphere.

But with the extra Greenhouse Gases trapping more heat in the lower atmosphere, there's less heat passing upwards through the upper atmosphere. This is because Greenhouse Gases are mostly in the lower atmosphere (0–10 km). So the heat that would have otherwise passed through the upper atmosphere and then into space is instead being 'caught' in the lower atmosphere and 'transmitted' back down to the surface. Hence, the upper atmosphere is cooling, while the lower atmosphere is warming.

We know this because our satellites (400 km and higher) have measured upper atmosphere temperatures for decades. The satellite measurements show a clear trend – there's less heat leaving Earth's surface and going into space.

The fact that the upper atmosphere is cooling while the lower atmosphere is heating is sometimes (incorrectly) used by some folk to 'prove' that climate science is all wrong! This is a debating trick designed to mislead rather than enlighten you.

LAND

Because we live on land (rather than on the oceans), the effects of Climate Change are so much more obvious. We're seeing rainfall patterns changing, deserts expanding, more frequent and intense heatwaves, snow cover receding, permafrost melting, glaciers melting, hurricanes and cyclones becoming more powerful, and so much more. Here are just a few examples.

Fires, heatwaves and drought – more intense and more often

Bushfires (called wildfires in the USA) are increasing in intensity and frequency. The 2019–2020 bushfires in Australia were the worst on record – covering some 300,000 km^2. In California, the 2020 wildfires burned more than 16,000 km^2 – doubling the previous record of 8,000 km^2 set in 2018.

Heatwaves are happening more often, and they're more powerful and deadly (see pages 124–125).

Drought risks are increasing across the globe.

In Australia, climate conditions are now so extreme that the Bureau of Meteorology has had to introduce a new colour (purple) for extra hot on the temperature maps, and our fire agencies have devised a new category (catastrophic) for bushfire risks. This is slightly less direct than the

US Embassy classification of ultra-extreme air pollution in Beijing – 'crazy bad'!

In 2020 in the USA, Phoenix, Arizona's daily temperature was 100°F (37.8°C) or higher for an unprecedented 144 days.

Storms are more intense, more often

Tropical cyclones (also called hurricanes or typhoons, depending on their location in the world) are becoming more powerful.

On 22 December 2020, *Scientific American* reported that 27 of the 30 storms that had occurred that year in the Atlantic had formed earlier than any year previously recorded.

Scientific American also wrote that 2020 was the first year in which 12 storms powerful enough to be given names hit the American landmass in a single season. Let me explain. First, not all tropical storms (the more powerful ones are called hurricanes in the USA) get official names – only the really powerful ones. Second, not all hurricanes actually hit the US landmass – some swing out to sea and dissipate there. Before 2020, the year with the most named storms making landfall was 1916, with nine hurricanes. But 2020 easily broke that record, with 12 named storms making landfall.

The list of increasingly stronger storms goes on and on.

Glaciers and permafrost are melting

Glaciers, which are masses of ice, sometimes have a 'glacial lake' into which the summer meltwater flows. Over the past 30 years, however, the volume of water in glacial lakes has increased by about 50% worldwide, because more glacial ice is melting.

Then there's permafrost – land that has remained at temperatures below freezing (0°C) for two or more years. I was astonished to find that as much as one-quarter of all the land area in the Northern Hemisphere is permafrost – some of it more than 1 km deep.

There are huge amounts of carbon locked up in the permafrost – about twice as much carbon as we humans have ever dumped into our atmosphere across all of human history. But the permafrost is now beginning to thaw. As it melts, this carbon is being released – mostly as carbon dioxide or methane. (Remember, methane is some 20–85 times more powerful as a Greenhouse Gas than carbon dioxide.) If these various Greenhouse Gases get to a dangerously high level, they will cause a massive runaway positive feedback loop of increased warming.

In the short term, nations with permafrost regions might have easier access to mine raw materials if the permafrost melts. Russia especially is about to benefit from this. But the long-term price is way too high.

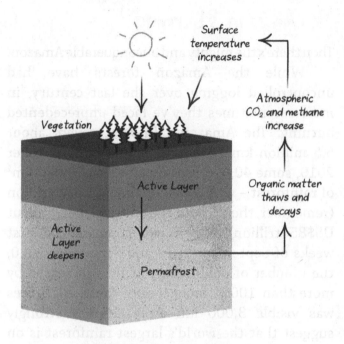

Permafrost is a major sink of terrestrial carbon. However, if Global Warming causes permafrost to thaw, the carbon dioxide and methane it has captured will be released into the atmosphere.
(Based on an illustration by the United Nations Environmental Program)

Rain (not snow) is falling in Alaska in unprecedented amounts. The five years from 2014 to 2019 were Alaska's wettest years in its century-long meteorological record. It turns out that 10 mm of rainfall can melt up to 7 mm of permafrost – which is very bad.

The Amazon is burning

Then there's the mighty and unconquerable Amazon.

While the Amazon forests have had uncontrolled logging over the last century, in more recent times they've faced unprecedented burning. The Amazon rainforest covers about 5.5 million km². Between January and October 2019, some 40,000 fires burnt some 10,000 km² of rainforest – with a cost of almost US$1 trillion (remember, the GDP for the entire planet is about US$85 trillion). When you compare the first weeks of September 2019 and September 2020, the number of deliberately lit fires increased by more than 100%. Smoke from these 2020 fires was visible 3,000 km away. Studies strongly suggest that the world's largest rainforest is on the brink of an irreversible 'dieback', which is another potential positive feedback loop.

The Amazon could turn into a landscape similar to the African savannahs or South American pampas. That single dieback event would release more than 500 billion tonnes of carbon dioxide. That's about one-third of all the carbon dioxide we've added to the atmosphere since the Industrial Revolution. That event would also be essentially irreversible.

The loss of the rainforest would have another effect. It would drastically reduce the rainfall in all the countries south of the Amazon.

Asian monsoon

More than 60% of the world's population lives in the Asian monsoon regions. The climate in those regions is complicated, due to the mix of different geography, local winds, ocean currents, etc. Monsoons have a wide influence, and are a major driver of global water cycles.

Projections? It's complicated, but there will likely be more very heavy rains – more so than in the past.

Arctic land

Think about the land areas in the Arctic.

Siberia has experienced unprecedented heatwaves, fires and permafrost melting (see pages 98–99).

Nunavut, in northern Canada, is now hotter than at any time in the last 115,000 years. The summer temperature in July 2020 was 5°C higher than the 30-year average. On 26 July 2020, a measuring station near the Milne Ice Shelf, the last existing Canadian ice sheet, recorded temperatures of 19.6°C. In late July, the ice shelf collapsed, losing 40% of its area in just two days.

Finally, let's look at Spitsbergen, the largest island in Norway's Svalbard archipelago. The administrative centre is Longyearbyen, which is located about 78°N, well and truly inside the Arctic Circle. The archipelago is home to about 3,000 people. The whole region has had unprecedented warming and flooding.

Average annual temperatures have risen by 4°C since the 1970s. However, winter temperatures have risen by more than 7°C. Unprecedented (there's that word again) summer temperatures of 21.7°C were measured in 2020. This has created some problems.

Problem 1. The previously solid permafrost is melting in Longyearbyen, so houses are sinking into the unstable ground. Some 250 dwellings will have to be torn down and replaced by buildings mounted on steel pillars driven deep into the permafrost.

Problem 2. The Svalbard Global Seed Vault in Longyearbyen was supposed to be an emergency, untouchable and permanent back-up for some 45,000 varieties of seeds from around the world. The reason it was built in Longyearbyen was because the permafrost was permanently (and naturally) colder than -6°C. But – you guessed it – global warming has caused the entrance to flood several times. A new entrance has been built, but it now has to be artificially frozen (with electricity) to avoid further erosion.

Problem 3. There's a coal mine just outside of Longyearbyen. Meltwater from a nearby glacier now floods the mine.

So, in the Arctic region, where temperatures are rising fastest from Climate Change, a coal mine is being flooded – with coal being one of the main worldwide contributors to Climate Change.

Is that called 'irony' or 'you reap what you sow'?

EARTH HAS BEEN TIPPED OFF ITS AXIS

Bizarrely, an accidental side effect of Climate Change is that we humans have managed to tip our entire planet off its spin axis. It's only by a tiny, tiny, tiny amount – but this is still a pretty astounding outcome!

After all, humans physically don't take up very much space, compared to the size of our planet. You could squish all of humanity into a cubic box that's 1 km on each side. That's very small compared to a ball some 12,700 kilometres across.

Given that there are relatively so few of us, it makes sense that most of Earth is completely untouched by humanity. We've made virtually no changes to the solid crust (5–70 km thick), apart from digging lots of holes and shifting stuff from here to there. For example, the deepest hole we've drilled is only 12.2 km deep, leaving another 6,359 km until we get to the centre of the planet. We've definitely done nothing to the gloopy molten rock of the mantle (the top 2,900 km immediately below the crust), and nothing to our planet's iron core. (Don't worry, I'll get to the North–South spin axis tipping really soon.)

Back on the surface, we humans have written all over the land part of our planet.

Forests have been replaced by plains, deserts or human structures. Rivers have been diverted and dammed, and roads criss-cross the surface.

Plus, there are those invisible 1,600 billion tonnes of carbon dioxide we've dumped into the atmosphere since the Industrial Revolution, which, as we now know, traps the equivalent of 400,000 Hiroshima bombs' worth of extra heat in the atmosphere each day.

But back to tipping the Earth. Surprisingly, we've managed to shift enough water from the Poles towards the Equator to change the tilt of the entire planet! (Again, only by a microscopically tiny amount.)

Here's what's happened.

The Earth's spin axis runs through the North and South Poles. But it doesn't run true – it wobbles a little. There's a bunch of reasons for this.

First, the Earth is not a perfect ball. It bulges a little at the Equator, thanks to the Earth's spin causing 'centrifugal force'. The diameter at the Equator is about 42.6 km larger than through the Poles.

Second, Earth's surface is a bit bumpy – while

the oceans are mostly smooth, mountains and valleys are not.

Third, Earth is not 'solid'. There's liquid oceans at the surface, then the 'gloopy' rock mantle reaching about halfway to the core, plus part of the core is liquid iron. So, our planet is a bit 'stretchy'.

Finally, the Earth's spin wobbles a little due to changes in water pressure on the ocean floor, changes in atmospheric pressure on the oceans and land, and other factors.

As a result, the North–South axis wobbles – in fact, there's a bunch of wobbles. One major wobble is the Chandler Wobble, discovered by the American astronomer Seth Carlo Chandler in 1891. Over about 433 days, the North Pole sweeps out a rough circle about 9 metres across.

But more importantly for our story, in addition to the Chandler Wobble, the North Pole has been drifting slowly south at about 6–7 cm per year.

Then, in 2005, some of the North Pole's motions unexpectedly changed.

First, the North Pole stopped heading for the Equator, and started moving east, parallel to the Equator.

Second, the movement of the North Pole accelerated to about 24 cm per year.

Finally, the Chandler Wobble reversed its phase.

What caused this?

Simple physics. For example, if you put a small weight onto a smoothly spinning bicycle wheel, it will start to wobble.

With Climate Change, the ice on land (mostly located away near the Poles, or at high altitude) began to melt, and eventually ended up in the oceans, at sea level. 'Centrifugal force' pushed this extra water towards the Equator. The redistributed water forced the previous 'spinning' of the Earth to change – in a very tiny way.

Back in 2014, the heat of Climate Change (you might remember, about 400,000 atom bombs' worth per day?) was melting at least 600 billion tonnes of land ice each year. This ice mostly came from Greenland, Antarctica and mountain glaciers. In 2020, that number was up to about 750 billion tonnes per year – or more.

On the one hand, this change in the 'wobble' of our Earth is so tiny that, of itself, it will have no effects. But I'm astonished that we insignificant humans can tip an entire planet off its axis at all.

It seems that, with Climate Change, we pushed our planet so hard, that it pushed back with a little wobbly of its own …

The next two chapters will look at the major effects of Climate Change in two areas – plants and animals, and people and the economy. But sometimes, these neat categories blend into each other – which is not surprising, because they are all interconnected.

6

HOT AND UNHAPPY PLANTS AND ANIMALS

The effects of Climate Change on nature

It's pretty obvious that changes in our environment ripple on and affect everything – bacteria, humans, and, yes, plants and animals. In general, living creatures have evolved to suit their local environment. While bacteria can produce a new generation every

20 minutes, animals take much longer, so this current episode of human-caused Climate Change is happening much faster than animals can evolve and adapt.

We could be here for years talking about plants and animals, but we don't have the time – they're feeling the effects now! So, I'll just give two short and general cases (plant timings and coral), and one longer specific case (pteropods), to show you how all this extra carbon dioxide can affect plants and animals.

In general, animals, insects and plants live most comfortably in their temperature zone. So, for example, imagine that they live on a mountainside, a little way down from the peak. Sure, if their local climate warms up over a period of time, they'll evolve to suit the warmer conditions – but this takes many generations. If the climate heats rapidly, they don't have time to evolve. In this case, all they can do is migrate a little higher up the mountain, into a cooler zone.

The problem is, once they reach the top of the mountain, if the local climate is still warming, they have nowhere left to go. Amazingly, we're already seeing the extinction of some animals, insects and plants due to this exact scenario.

ANIMAL AND PLANT TIMINGS AND LINKAGES CAN GET OUT OF WHACK

It's one thing for biologists to know that the heating of the Earth's biosphere (where all the plants and animals live) would force some species to migrate. But it's another thing to discover just how species would move and how quickly.

So, when local climate changes, the local species have only a few choices – change themselves via evolution, tolerate the changed climate, go extinct, or migrate.

Usually, with Global Warming, the species on land will migrate towards the Poles or to cooler higher elevations to get away from the heat – but sometimes they will migrate downhill to stay in a more humid climate. In the oceans, the species will usually migrate Poleward to the cooler waters, but sometimes they migrate deeper to the cold water at the greater depths.

One unexpected result is that a bunch of new hybrid species are evolving, because different species have migrated into the same area and bred with each other. We have found hybrid species of toads, butterflies, trout, sharks and bears.

On the southeast coast of Australia, the sea urchin has expanded its range enormously with rising water temperatures. The result has been a loss of large seaweed, which in turn has led to much-reduced catches for popular fish.

In north Alaska, deep in the Arctic, Climate Change has altered the diet of the indigenous peoples. Their traditional foods, such as seals, have become harder to get because the sea ice is melting sooner and for longer. Now they eat moose and snowshoe hares.

The reason is that back in the 1800s, trees such as the alder and the flowering willow grew to only a metre in height. But with Global Warming, not only was the local temperature higher, but the growing seasons got longer. So now the trees are two metres tall, high enough to poke out above the snow – and for the moose to easily eat. So moose began migrating northward in the 1900s, shortly followed by another shrub-eating animal, the snowshoe hare.

In Sweden, the so-called Lake of the Pine Trees no longer has any pines – because of Global Warming, it is now surrounded by birch.

At the moment, much of the European forest areas are covered with valuable trees such as the Norway spruce. But with Climate Change, about 34% of these areas will change and become suitable only for less valuable trees such as Mediterranean oak.

In some cases, intertwined ecological relationships that involve precise timing have come apart.

In the Arctic, snow is now melting weeks earlier, so insects are hatching much earlier. This is a problem for a migrating shorebird, the red knot, that flies from the tropics to the Arctic in springtime to feed on insects and breed. Nowadays, the insects have already hatched and moved out before the baby red knot chicks hatch – so the chicks can't eat their regular food. There is another source of food – molluscs on the sandy beaches in the Arctic in spring. Unfortunately, the beaks of the red knot chicks are too small to eat them.

In Japan, the bumblebees are emerging after the herb *Corydalis ambigua* has flowered earlier than before – so there are fewer seeds produced, and the herb is fading away.

In West Greenland, thanks to another timing mismatch, the pregnant caribous can't eat enough plants, so the death rate of the newborn caribou is increasing.

Diseases and pests also migrate with the shifting climate bands.

In South Africa, diamondback moths are now attacking cauliflower, cabbages and kale in areas where they previously did not, while in Latin America various fungi and pests are attacking coffee plants in locations that had previously been disease free. In France, migrating diseases have harmed crops, such as lavender, olives and wine grapes.

While the overall trend on the flora and fauna has been negative, there have been a few patches of positive outcomes. The Atlantic mackerel have migrated northwards, so that the catch around Iceland has increased from 1700 tonnes in 2006 to 120,000 tonnes just four years later in 2010.

In part of the tropical Pacific, the skipjack tuna is expected to increase in numbers. This will offset the overall decrease in fish numbers, due to degraded coral reefs.

Human societies are deeply linked to the life and climate around them. Many ancient civilisations have collapsed because of changes to their local biosphere – such as the Indus Valley Civilisation of Pakistan, which collapsed about 3,700 years ago due to a mega-drought.

Survival is not just for the fittest – it also depends on still fitting in.

CORAL, FISH AND WARMER OCEANS

Coral is an animal. (Yes, even though it's glued in place and can't move once it settles down, it's still an animal.)

Mass bleaching of coral reefs was uncommon before the 1960s. Now, thanks to the oceans warming rapidly, bleaching is threatening most coral reefs globally. Sure, coral can adapt over many generations, but ocean temperatures are rising so rapidly that the coral can't evolve fast enough. So the coral dies. On the Great Barrier Reef, mass-bleaching events are becoming more frequent – 1998 was the first one, followed by four others in 2002, 2016, 2017 and 2020.

Not surprisingly, coral is linked to fish. Even though coral reefs cover only 0.1% of the ocean's surface, 25% of all marine species are born or live there.

It's pretty obvious that less coral reef means fewer baby fish. And that means trouble for the billions of people living in Asia – because about one-quarter of all their fish comes from coral reefs.

So, Global Warming is savagely disrupting fish stocks – and further down the line, we'll have less fish to eat.

PTEROPODS AND OCEAN ACIDIFICATION

Unfortunately, carbon dioxide is making the oceans more acidic. Unless we stop this, our future marine food supplies will be threatened.

So, how does carbon dioxide make the oceans more acidic?

$$CO_2 + H_2O \longleftrightarrow H_2CO_3$$
Carbon dioxide + water $\longleftrightarrow$ carbonic acid

About one-third of all the 1,600 billion tonnes of carbon dioxide we've dumped into our atmosphere since 1750 has dissolved into the waters of our oceans, lakes and rivers.

The good thing about this carbon dioxide leaving the atmosphere is that it takes longer for the air temperature to get hotter.

The bad thing is that, as the carbon dioxide dissolves, it makes the water slightly more acidic. (To be specific, it shifts the oceans from slightly alkaline to slightly less alkaline. In other words, the pH moves a little towards the acid end of the pH scale, where the lower numbers are.)

Before we go any further, you need to know that the pH scale ranges from 0 (incredibly acidic) to 7 (neutral, neither acidic nor alkaline) to 14 (incredibly alkaline). In preindustrial times, the pH of the oceans was about 8.2. By 2020, it had shifted slightly in the 'acid' direction to about 8.1 (a drop in pH of 0.12).

Now, here are two important points.

First, you wouldn't notice this slight acidification with your skin. You can dip your foot into the surf at your local beach and it won't feel any different.

Second, a change in pH of 0.12 sounds very small. But (talking in the language of mathematics, sorry) the pH scale is logarithmic, not linear. This means that this 'tiny' change in pH of 0.12 towards more acidic reflects an enormous 30% increase in the number of available hydrogen ions (sorry about the simplified chemistry, but hydrogen ions cause acidity). These hydrogen ions then combine with 30% more carbonate ions. Depending on the local pH, the following reaction can go in either direction.

$$H_2CO_3 \longleftrightarrow 2H^+ + CO_3^{2-}$$

These carbonate ions used to be available for creatures in the ocean to make shells or skeletons. Such lovely creatures include crabs, mussels, krill, sea urchins, plankton, coral and the unfortunate hero of my story – the pteropod.

Christopher Monckton's false claims about acid in the ocean

The English Climate Change denialist Lord Monckton has totally dismissed ocean acidification. 'Our harmless emissions of trifling quantities of carbon dioxide cannot possibly acidify the oceans,' he wrote. 'Paper after paper after learned paper in the peer-reviewed literature makes that quite plain ... There is no basis for imagining that we can acidify the oceans to any extent large enough to be measured even by the most sensitive instruments.'

He is so incorrect!

Monckton claims that the 'quantities' of our carbon dioxide emissions are 'trifling'. They're not. In 2018, they were about 37 billion tonnes – which is roughly equal to the mass of all of the dirt and rocks shifted by all of the rivers on our planet.

Monckton also wrongly claims that the oceans are not more acidic. The opposite is true – the oceans are more acidic than they were a few centuries ago. How do we know? We measured it – a drop in pH of 0.12.

Monckton then claims that most scientific peer-reviewed literature backs up his assertion. Again, he's wrong. The scientific peer-reviewed literature discusses the acidification of the oceans and there's overwhelming agreement that the oceans are getting more acidic.

Monckton's last sentence is also wrong. (At least, he's consistent!) He claims that even our most sensitive instruments cannot measure any change in the pH of the world's oceans. But for about A$200, you can buy a pH meter that is accurate, not to 0.1 pH, but ten times better – 0.01 pH! That's quick and cheap as chips, delivered to your house in a week.

Meet the pteropod

Since the mid-1970s, when scientists started thinking seriously about Global Warming, they were also considering the possibility that ocean acidification could affect sea creatures. They didn't think it would happen for a long time. But they didn't foresee just how much carbon dioxide we'd dump into the atmosphere as a result of burning fossil fuels.

In 2012, a peer-reviewed report in the scientific literature said that ocean acidification had already badly affected the pteropods in the Southern Ocean.

The pteropod is a small sea snail, about a centimetre across, that lives for a few years. It swims freely in the oceans, using wings like a little butterfly. But today, the pteropod can no longer make a perfect shell. Instead, thanks to ocean acidification that is mopping up many of the available carbonate ions that it needs, the pteropod's shell is malformed and fragile.

You might think, 'I've never heard of a pteropod. So what if it can't make its shell anymore? It can't possibly affect me, surely.'

But the pteropod affects humans in two problematic ways – carbon flux and food.

Disturbance in the flow

The first problem is that malformed pteropod

shells will cause a disturbance in the carbon flux of our planet's surface.

Confused? Sure. Let me introduce you to the concept of 'carbon flux', a fancy word meaning 'flow'.

I'll start with the carbon atoms in the Purple King Climbing Beans in my garden. I eat some of the beans (they grow so quickly and use up so little garden space – I love them to pieces). My body digests the carbon atoms, which then leave my body through my major excretory organ – my mouth. They exit married to a few oxygen atoms in the form of carbon dioxide.

When I breathe in, air with carbon dioxide at 0.04% concentration floods into my lungs. But the air that I breathe out carries some 4% carbon dioxide – 100 times more. By random good luck, some of these carbon dioxide molecules might drift near the tomato plants in my garden, and get incorporated into the tiny delicious cherry tomatoes they carry.

So there has been a flux (or flow) of carbon atoms from the Purple King Climbing Beans, into my mouth, then to my tummy and out of my mouth – then over into my tomatoes.

I eat the tomatoes. After a while, the carbon atoms end up back out in the atmosphere. Maybe next time, some will end up in the frangipani tree in my front garden.

So now, you get the idea of carbon atoms

continually moving around in the biosphere. This is Carbon Atom Flux – carbon atoms moving around continually in the biosphere.

Back to the tiny pteropod. How many carbon atoms travel through it? Maybe, you'd guess, not many? Maybe one in every million trillion?

No!

One in every twenty or so carbon atoms in the biosphere travels through a pteropod.

What if the pteropods die out? What would happen to all that carbon that they used to process? Where would it go? Would it stay in the oceans to make them even more acidic?

Disturbance in the food

We face another problem if we lose the pteropod. In the northern Pacific Ocean, pteropods can make up as much as 60% of the food eaten by juvenile pink salmon. We are potentially interfering with our future smoked salmon (which I love, but that's not the big issue).

Acidification interfering with the pteropod is just the start. There are many other consequences.

Acidification will interfere with the transfer of many other essential elements in the ocean, as they continuously move into and out of living creatures. These elements include iron, chromium, zinc and so on.

Beginning around 2007, we have seen the collapse of commercial oyster hatcheries in

Oregon, USA. This is because the oyster larvae are badly affected by acidification at a critical and early life stage. It's tragically obvious – no oyster larvae mean no adult oysters.

THE BIG PICTURE

Over the past billion years – long before we humans evolved onto the scene – the pH of the oceans has changed (turning either more acidic or alkaline) many times.

But now, the oceans are turning more acidic more quickly than at any other time seen in the geological record over the past 300 million years. Furthermore, the oceans have not been this acidic (pH 8.1) at any time in the last 15 million years. Sure, ocean creatures will evolve to handle the new pH levels – but this takes thousands of years, not decades.

Ocean Acidification is the Evil Twin of Global Warming.

The web of life is truly interlinked. When we change the physical environment, the biosphere changes for all living creatures.

But don't freak out – Chapter 8 carries messages of Good Hope.

In the first *Star Wars* movie, Princess Leia had Obi Wan-Kenobi as her only hope. We've got a whole lot more hope than that!

7

HOT AND UNHAPPY PEOPLE

The effects of Climate Change on us

Finally, we get to the direct and personal effects of Climate Change on you and me. It's already hitting our health – physical, mental, emotional, etc. – and our wallets. Let me give you a quick snapshot of what's already happening across the globe, and what is projected to happen.

MIGRATION

We're seeing increases in human migration and conflicts due to Global Warming.

It's easy to see why people are migrating away from coastal flooding (following sea-level rise). They are already leaving parts of Florida, Asia (China, Bangladesh, Vietnam) and many Pacific islands.

Migration is also triggered by the loss of farmland and crop failures. These so-called 'Climate Refugees' were in Senegal, Mozambique, Central America and many other places. Previously fertile land can turn into desert, or fresh water needed for growing crops can disappear.

Migration is also occurring as a result of mountain glaciers delivering less meltwater in summer. This is due to reduced precipitation in winter, which means less snow and ice formations. Surprisingly, the drainage basins of mountain glaciers cover about one-quarter of the global land surface area (excluding Greenland and Antarctica), and are home to one-third of the world's population. So any reduction in glacial meltwater can affect many people. The greatest glacial shrinkages (as measured from 2003 to 2009) were in North America (110 billion tonnes per year), the Himalaya region (26 billion tonnes per year) and South America (29 billion tonnes per year). The reduced irrigation means that some areas can no longer be used for agriculture and, so, are being abandoned.

DISEASE

As climate zones move away from the Equator, many diseases – including malaria, dengue fever, Ross River virus and more – are following.

Consider malaria, which infects more than 200 million people each year. Malaria is carried by the *Anopheles* mosquito, which is migrating both towards the Poles and up the mountains away from the heat. This larger range for the mozzie will increase the number of infections. We are already seeing this happening in the higher mountain slopes of Colombia and Ethiopia, which previously had much lower incidences of malaria.

In the Komi Republic (to the west of the Ural Mountains in Russia), an encephalitis (infection of the brain) carried by ticks has increased by a factor of six.

In northern Sweden, over the past 30 years, a disease of humans called Rabbit Fever (carried by insects and animals) has increased by a factor of ten.

Some fungi seem to benefit from the higher temperatures – at the expense of humans. The yeast *Candida auris*, which was discovered only as recently as 2009, can cause fatal blood infections, especially in nursing homes. (It does not cause thrush.) Ordinary room temperatures usually kill *Candida auris* before it can infect people. But higher temperatures (due to Climate Change) are forcing it to evolve rapidly, and there are now several strains that can easily infect us.

In early November 2020, a group of more than 700 medical doctors wrote to the Australian prime minister, accusing the federal government of failing to protect Australians from the impacts of Climate Change.

Some of the signatories were Professor Nick Talley (editor-in-chief of the *Medical Journal of Australia*), Dr Clare Skinner (President of the Australasian College of Emergency Medicine), and Associate Professor Peter Sainsbury (School of Public Health, University of Sydney).

They wrote: 'Public health is inextricably linked to climate health. Climate damage is here now and is killing people ... Australians are already seeing higher rates of respiratory illness, diarrhoea and morbidity requiring hospital admission during hot days, and higher rates of suicide in rural areas during drought years. The burning of fossil fuels ... is linked to the premature deaths of 3,000 Australians each year.'

DIRECT IMMEDIATE DEATHS

Heatwaves caused by Climate Change have killed tens of thousands of people.

In the terrible 'Black Saturday' disaster of 2009 in Victoria, flames, heat and smoke killed

173 people directly. But the heatwaves before, during and after Black Saturday, which had been caused by Climate Change, killed more than twice as many – 374.

The 2003 European heatwave killed about 70,000 people. The 2010 Russian heatwave killed about 55,000 people.

In the 1960s, in any given year, heatwaves covered only about 1% of our planet's land area. This reached 5% by 2010, and by 2020 it was about 10%. It's projected to reach 20% by 2040.

Heatwaves are especially lethal when combined with very high humidity. In this situation, you cannot easily cool down through your sweat evaporating. When high-humidity areas get too hot, they will be effectively uninhabitable.

THE COST

Take Queensland ...

Climate change is already having a major economic impact on societies. But this impact will increase, especially if people don't act quickly to reduce carbon dioxide emissions and their effects, and to 'pivot' to new industries.

Take the State of Queensland. At the moment, it's not preparing for the future. Currently, Queensland depends heavily on:

- coal mining (an industry in global decline, which is becoming a stranded asset)
- agriculture (which is subject to increasing droughts)
- tourism (especially to the Great Barrier Reef, which is in swift decline, and to rainforests, which have burned for the first time in recorded history).

The Swedish Central Bank *is* preparing for the future, however; it has taken a long-term view and voted with its money. Having assessed Queensland as being susceptible to financial calamity, it has sold all its Queensland Government Bonds.

Disaster costs

According to *Scientific American*, in 2020, 'the US has had 16 natural disasters (including wildfires, hurricanes, tornadoes and drought) that each caused at least US$1 billion in damage'.

Over the past 20 years, 20 disasters (caused by Climate Change) have inflicted damage costing more than 10% of the GDP of the unfortunate countries that experienced them.

The tiny Caribbean country of Dominica has been especially unlucky. In 2015, Tropical Storm Erika caused damage equal to 90% of Dominica's GDP. Then in 2017, Hurricane Maria inflicted damage equal to 226% of Dominica's GDP.

Costs of the permafrost thawing

Dr Gail Whiteman, currently professor of sustainability at Exeter and Yale universities, looked at the release of thawing permafrost in just one tiny area – the East Siberian Sea of northern Russia. The cost, she said, has 'an average global price tag of $60 trillion … comparable to the size of the world economy in 2012 (about $70 trillion)'.

There's a huge amount of methane gas under the ocean floor there. With Global Warming, it could be released from the permafrost – either slowly over half a century, or much faster. The release of this amount of Greenhouse Gas (and remember, methane is some 20–85 times worse than carbon dioxide) would accelerate Climate Change enormously, and set off the release of even more methane – a classic positive feedback loop (see pages 84–85). Unfortunately, about 80% of the costs would be carried by poorer economies and developing countries in Africa, South America and Asia, in the form of extreme weather and heat stress, flooding and droughts, agricultural production, and so on.

On 20 June 2020, a record high temperature of 38°C was reported in the remote Siberian town of Verkhoyansk. If verified, this is the hottest temperature ever recorded inside the Arctic Circle. In 2020, Siberia was the most abnormally hot place on the planet. The permafrost should have been sub-zero in temperature, or at least wet and boggy – not dried out enough to burn. This is a very scary example of how fast the climate is changing in the Arctic, and how extreme these changes are. And this is also just another example of the destruction caused by Climate Change (and yes, 30 years after the climatologists showed it was happening, the fossil fuel companies are still spending big on funding denialist campaigns).

The year 2020 (not an El Niño year) shared hottest-ever year with 2016 (an El Niño year). And the six hottest years ever on record were 2015 to 2020. The wildfires seen raging across Siberia in 2019 and 2020 were the worst on record.

So how do we measure the cost of Siberian wildfires – especially when the 2020 Siberian fires emitted so much carbon dioxide, due to burning over peatland – before the actual hard costs come in? After all, it's very difficult to make reliable predictions, especially about the future. We can't estimate them exactly, but the economists are telling us that they will be enormous.

Costs of rising sea levels

Currently, more than 600 million people live in coastal zones less than 10 metres above the mean sea level. Under a Business As Usual scenario, projections for the year 2100 are grim. The scenarios vary enormously, depending on what we humans decide to do. At one extreme, we increase our greenhouse emissions, but, at the other extreme, we reduce them to zero and start pulling carbon dioxide out of the atmosphere.

The term 'a once-in-100-years event', which is used for certain extreme natural disasters such as floods, droughts, bushfires and cyclones, is rapidly becoming meaningless, because these events are becoming more commonplace.

Let's look at the year 2100, under the current high-emissions scenario. The previous once-in-100-years coastal floodings will happen every 10 years. And they will affect 48% more land area than they do today. These floods wil affect much of the world's population and will threaten assets worth 12–20% of global GDP.

Surely, we should act now to protect some US$14 trillion of our global assets, out of a total global GDP of US$85 trillion?

Meanwhile, we're paying billions to Big Fossil Fuel

While we're mentioning protecting our money, let's not forget the subsidies (that is, free money, with no obligation to pay anything back) that Big Fossil Fuel receives.

The International Monetary Fund report that, globally, governments paid out about US$5.2 trillion in subsidies to Big Fossil Fuel in 2017.

By itself, that was 6.5% of global GDP, or about 8% of all government revenues averaged out across the planet.

There are many different ways of working out the costs and effects of Climate Change. But they all have one thing in common – the cost is huge (at least several per cent of the total Gross World Product) and it will be *even higher* in the future. Just as prevention is better than a cure, it will be much cheaper to fix the problem now.

This situation is remarkably similar to the financials of cigarettes. Governments collect a small amount of revenue today from taxes, exports and so on – but the cost to the community is five times larger, and it is paid in the future.

8

WHAT'S THE FIX?

There's heaps we can do to repair and reverse Climate Change

I don't want to demonise poor old carbon. This lovely little atom is innocent. It's built into every single organic chemical and is fundamental to life. It's what we currently do with carbon that's the problem. The good news is we can both stop and even reverse Climate Change, by using the science and technology we already have — though breakthroughs will certainly occur and they will make the job even easier and faster.

So why haven't we fixed and reversed Climate Change already? The answer is, a massive disinformation campaign by Big Fossil Fuel.

And how can we fix and reverse it quickly? Remarkably easily.

WHOSE CARBON FOOTPRINT?

Now, if you've been following the science of Climate Change, you'll have come across the term 'carbon footprint'. This refers to the total amount of Greenhouse Gas emissions created by an individual, product, event, service or organisation, or country.

What you probably don't know is that the Big Fossil Fuel company BP popularised the term back around 2005. It was a cunning marketing ploy to shift responsibility for Greenhouse Gas emissions from Big Fossil Fuel companies to individuals. The hidden backstory is that the majority of an individual's carbon footprint is set up by the society they live in.

Sure, it's good for everybody (citizens, companies and governments) to reduce their emissions as much as possible. It's a feel-good thing that resonates with all branches of society.

Unfortunately, however, the actions of individuals can reduce global emissions only slightly. In reality, making significant reductions depends on a major shift at the level of industry and government to move away from emitting Greenhouse Gases.

Per person, the global average 'carbon footprint' is about 5 tonnes. Mind you, for countries such as Australia and the USA, it's up at about 20–25 tonnes per person – so that means there are vast numbers of people around the world who emit a lot less. (As an aside, if you include all the fossil fuels that Australia exports around the world, our carbon footprint rises to about 70 tonnes per person. Those exports bring Australia's emissions to about 1.8 billion tonnes of carbon dioxide per year, or about 3.6% of global annual emissions. And remember, we're only 0.3% of the world's population.)

In Western societies, even a homeless person has a carbon footprint of about 8.5 tonnes of carbon dioxide per year – which is 70% higher than the global average. This is because our Western societies are deeply married to fossil fuels.

In fact, almost 50% of global greenhouse emissions come from just 10% of the people – the wealthiest 10%. Furthermore, the richest 1% of the world's population are responsible for more than twice as many carbon emissions as the 3.8 billion people who make up the poorer half of humanity.

So how much can one Australian do?

Well, the biggest difference is having one fewer child in the family. After all, each Australian has a carbon footprint of 20–25 tonnes per year.

Everything else has a much smaller impact:

- Installing a smallish 3-kW solar power plant on your roof (in a region with 150–250 days of sunshine each year) saves about 3.3 tonnes of carbon dioxide compared to electricity generated by burning fossil fuels.
- Going without a car saves 2.5 tonnes of carbon dioxide per year.
- Shifting to an entirely plant-based diet will save about 0.75 tonnes per year.
- Washing your clothes in cold water and drying them on a clothesline will each save about 0.25 tonnes per year.

What about a pandemic? How does that affect our greenhouse emissions? Well, when COVID-19 hit in early 2020, there were rolling lockdowns worldwide. People worked from home and barely left their suburbs. Globally, transport and production dropped enormously, and economies suffered. And yet, global carbon dioxide emissions for 2020 dropped by only 6.4% (about 2.3 billion tonnes) compared to 2019, with the USA having the biggest reduction.

Big Fossil Fuel's carbon footprint

The reality is, fixing Global Warming will take more than the actions of individuals, or even a global pandemic.

Instead, those who really need to change are the fossil fuel companies and the governments that continue to support them, because the best way to reduce the world's carbon footprint would be to shift our energy production from fossil fuels to renewables.

Unfortunately, the main reason we're in our current poor situation is due to the actions of Big Fossil Fuel and the remarkably compliant government bodies they have successfully lobbied for so long. Against these big actors, private citizens are powerless – like a little rowing boat towed behind a giant supertanker.

As we saw in Chapter 1, Big Fossil Fuel moved from *funding research* into Climate Change back in the 1970s to *denying* Climate Change in the 1990s while, at the same time, funding massive disinformation campaigns and lobbying governments to keep burning fossil fuels. The direct result is that 19 of the past 20 years have been the warmest years on record.

Blaming consumers for causing Global Warming is like blaming smokers for getting addicted to tobacco products. For decades, Big Tobacco knew cigarettes were addictive and carcinogenic and killed people – and yet, they continued to sell them.

But victim blaming is basically what BP, one of the world's largest fossil fuel companies, did with their very successful advertising campaign about the carbon footprint back around 2005.

A classic case of 'victim blaming'

Blaming individuals for causing carbon emissions is like blaming people who own refrigerators for using them – and thus causing depletion of the Earth's protective ozone layer. This ozone depletion is exactly what happened with chlorofluorocarbons back in the 1980s. Chlorofluorocarbons (CFCs) are chemicals that attack and destroy the high-altitude ozone layer. When CFCs were introduced in the 1930s as a refrigerant gas, nobody knew how dangerous they were. They were seen as a wonderful solution. If they leaked out of your fridge, they didn't poison you, and they didn't burn or explode. But half a century later, scientists discovered that CFCs were damaging the ozone layer, which protects life on Earth by absorbing some of the Sun's harmful UV radiation.

The several dozen companies making CFCs tried to deny this fact. They also claimed that there were no possible alternatives to CFCs as a refrigerant gas. But the companies were wrong or telling mistruths – and the scientists were right. The governments of the world banned CFCs via the 1987 Montreal Protocol.

Imagine if those CFC-making companies had been able to stop the Montreal Protocol and continued to pump out CFCs, destroying the ozone layer in the process and exposing us to more UV radiation? That would be on a par with running a massive disinformation campaign telling us that we were the cause and to use fridges less often!

BP – Big Polluter

BP has one of the worst environmental and safety records of any Big Fossil Fuel company. It was responsible for the largest-ever accidental release of oil into marine waters, with the 2010 *Deepwater Horizon* oil spill in the Gulf of Mexico. (The company eventually pleaded guilty to 11 felony counts of manslaughter, 2 misdemeanours, and 1 felony count of lying to US Congress about how much oil was gushing out of the blown-out underwater well. Ultimately, BP had to pay more than US$65 billion in fines and claims.)

Back in 2001, in an effort to polish their image, BP had rebranded by renaming itself 'beyond petroleum' (spelled in lowercase because, according to one of the people in the rebranding exercise, 'lowercase is cool'). It also gave itself a nice bright yellow-and-green sunburst logo.

With great fanfare, BP unveiled their 'carbon footprint calculator' in 2004, which showed ordinary citizens how their own activities were contributing to Global Warming. The subtext was that the real cause of Global Warming was families using too much electricity, flying on planes, and going to and from work. BP also told us that it's 'time to go on a low-carbon diet'.

And yet, BP is still producing virtually as much oil and gas today (3.8 million barrels per day) as it did in 2005 (4 million barrels per day). In 2019, BP made its largest purchase since 2002, buying gas and oil reserves in West Texas. According to BP, this was to give them 'a strong position in one of the world's hottest oil patches'.

How much of its budget does BP invest in renewable energies? Over the last decade, it's been a microscopic 1% or so. They don't sound very serious about a low-carbon diet for themselves, do they?

90 COMPANIES GENERATE TWO-THIRDS OF GLOBAL EMISSIONS

Two-thirds of the world's carbon dioxide and methane emissions are generated by just 90 major industrial carbon producers. Almost certainly, never before in human history have so few been responsible for so much.

Let's zoom in on these companies. Just 20 of them are directly linked to more than one-third of all modern-era Greenhouse Gas emissions. As Michael Mann, the climate scientist who came up with the famous 'Hockey Stick Curve' of Global Warming (see page 18), said: 'The great tragedy of the climate crisis is that seven-and-a-half billion people must pay the price – in the form of a degraded planet – so that a couple of dozen big companies can continue to make record profits.'

The Big Fossil Fuel companies have successfully slowed the transition from fossil fuel to renewables for our essential energy. They have done this for the last three decades. Fossil fuels are the major source of Global Warming. And yet, we citizens have to live in a world still mostly powered by fossil fuels.

According to Dr Benjamin Franta, a physicist who also researches law and the history of science at Stanford Law School, the BP carbon footprint campaign was 'one of the most successful, deceptive "public relations" campaigns ever'.

So yes, let's lower our personal carbon footprints. But we need to recognise that 'us' means *all* of us – individuals, governments and, yes, really big companies. And perhaps the biggest change we individuals can make is to switch to a superannuation fund that doesn't invest in fossil fuel companies. The other thing we can change is who we vote for.

With the hole in the ozone layer, we took quick and decisive action. In 1973, scientists discovered that CFCs could damage the ozone layer. Those scientists won a Nobel Prize. In 1985, data confirmed that ozone depletion was actually happening. Just two years later, in 1987, CFCs were banned.

With Climate Change, in about 1990, scientists confirmed that humans were causing it. Three decades later, we're still waiting for action from governments and businesses.

WHAT CAN WE DO?

Obviously, private citizens and responsible corporations can't make all the changes on their own. Like the Montreal Protocol that successfully banned CFCs, there has to be government action, and different governments will have to work together. But that doesn't have to take decades.

Governments can act quickly if they wish. They've certainly done it before. Take the US government and Pearl Harbor, for instance.

Move to a war(like) footing: the story of the B-24

On 7 December 1941, Japan bombed Pearl Harbor and, suddenly, the USA and Japan were officially at war. Until that point, the USA had been neutral, and not a combatant in World War II.

Immediately, the USA moved to a 'war footing'. Within one month of the bombing of Pearl Harbor, it ordered the entire American car industry to stop making civilian vehicles and switch over to military production.

Up until that time, the US aircraft industry had produced about 3,000 planes in total. But during the following four years, US car manufacturers built about 300,000 planes. This gave them an overwhelming technological advantage. For the Americans, the war was won as much by the machine shop as by the machine gun.

As an example of what we humans can do when we put our minds to it, take the American heavy bomber – the B-24 Liberator. This is a huge plane (20 metres by 34 metres) weighing up to 30 tonnes and with a crew capacity of 11. Thanks to its four enormous supercharged 1,200 horsepower engines, it has a top speed of 478 km per hour, and a range of nearly 2,500 km.

Compared to a car, which had 15,000 separate components, the B-24 had some 500,000 separate components, mostly of high-tech materials. Furthermore, every component had to be manufactured, then assembled and fitted in place – to much tighter tolerances than in an automobile.

Even so, just one Ford factory alone (Willow Run in Michigan) could pump out these highly complex machines at the rate of not one per month – but one per hour!

Looking up one of the assembly lines at Ford's big Willow Run plant, 1942–3.
(US Library of Congress)

How did they do that?

Step 1: Ford got a few B-24s and carefully broke them down into their 500,000 separate components. Then, more than 200 people spent the best part of a year drafting some 30,000 blueprints – which took up enough space to fill two shipping containers.

Step 2: From scratch and on virgin ground, Ford built the largest single-storey building in the world. It was about a kilometre long and about one-third of a kilometre across.

Step 3: Ford gathered together a huge workforce. More than 40,000 people were involved, including people with dwarfism who had been specially selected for their shorter size, so they could crawl inside certain parts of the plane where a taller person could not. Suddenly, workforce minorities (the short-statured, women and many ethnic groups) were able to get not just skilled work, but also get the same full wage as their white male colleagues.

This is what being on a 'war footing' means – the ability to pump out one B-24 Liberator heavy bomber per hour in response to an urgent need.

Shifting Climate Change into reverse

If we were to go on a war(like) footing, we could easily stop, and then reverse, Climate Change, returning Greenhouse Gas levels to what they were in the mid- to late-20th century.

Prevention is better than the cure, however, so the sooner we start, the better and cheaper it will be.

Change is always a little messy. But the current and future dangers and costs from Climate Change are truly horrendous. We *have* to change. The transitions may be messy – but that will be only for a short time, and the long-term results will be so much better. We can think of this in two parts.

Part 1: We need to reduce greenhouse emissions as much as possible, and as quickly as possible.

Part 2: Simultaneously, we need to reverse Global Warming.

Let's start with the first part.

PART 1: REDUCE EMISSIONS

Reducing emissions may seem like an impossibly huge task, initially. But it's a lot easier if you break the emissions down into a handful of major sources, or 'sectors'.

There are many different ways to subdivide the sectors of the world's economies in terms of greenhouse emissions. The one we used in chapter 4 (see page 72) is pretty good:

- Electricity and heat – 30%
- Transportation – 16%
- Manufacturing and construction – 12%
- Agriculture and livestock – 11%
- Other – 8%
- Land use and forestry – 6%
- Fugitive emissions – 6%
- Industrial processes – 5%
- Buildings – 5%

Here's what we can do to tackle several of these sectors.

Electricity and heat – from 30% emissions to zero in ten years

Some 30% of worldwide global greenhouse emissions directly relate to generating heat and electricity. And yet it would be easy to reduce these emissions to virtually zero within ten years.

In fact, about a decade ago, a think tank called Beyond Zero Emissions (bze.org.au), along with the University of Melbourne, released a document called the *Zero Carbon Australia Stationary Energy Plan Synopsis* that showed how Australia's electricity supply could transition from fossil fuels to 100% renewables within ten years. If put into action, the plan would have provided over 100,000 jobs in Australia and, overwhelmingly, these jobs would have been local. It also would have set up several high-tech industries in Australia, which could have exported this technology overseas. This plan could have provided rock-solid baseload power with two lovely advantages – better health within the Australian population from reduced emissions, and cheaper electricity. Yes, it turns out that, if you do not have to pay for the fuel (sunlight and wind are free), electricity is 60–70% cheaper.

If that plan had been adopted by our federal government back in 2010, Australia would have been one of the first countries on the planet to go zero carbon for electricity. Instead, the government has taken the opposite pathway – and worked hard to support the failing fossil fuel industry and weaken electricity generation from renewables.

Nevertheless, the Australian Capital Territory runs entirely on renewable electricity. There have been several occasions when renewables have provided more than half of Australia's electricity – for several hours at a time. Imagine how often this would happen if the Australian federal government recognised the urgent need to combat Climate Change even five years ago.

Beyond Zero Emissions continue to research and develop what they call their 'Zero Carbon Australia' plans, which we could start putting into action tomorrow. All we need is government and business to act.

Transport – from 16% of global emissions to zero in 15 years

About 16% of global greenhouse emissions relate to transport by road, sea and air.

Most short-haul transport vehicles, such as cars, buses, trucks and ferries, can be powered by electricity from batteries, or overhead wires for trains.

However, to provide clean power for long-haul transport (via sea and air), which accounts for about 5% of global Greenhouse Gas emissions, we'll need hydrogen power. Currently, we burn petrol, diesel and kerosene for transport. We've learned how to handle them safely, and we can do the same with hydrogen.

Remember the equation?

$$C + O_2 \rightarrow CO_2 + heat$$

We can burn hydrogen (instead of carbon) to get the heat we want:

$$2H_2 + O_2 \rightarrow 2H_2O + heat$$

Instead of getting a Greenhouse Gas that does not easily leave the biosphere, we get water that blends back into the biosphere very quickly. In some cases, we can do the reaction very slowly inside what is called a 'fuel cell'. A fuel cell is a box that gives you electricity so long as you keep feeding it the right fuel, such as hydrogen. It also has no moving parts, apart from the hydrogen and oxygen coming in, and the water coming out.

Hydrogen is the lightest element. The straight physics of this means that, volume for volume, liquid hydrogen carries only about one-quarter of the energy of kerosene or diesel.

Shipping is no problem, because ships have lots of spare room. Aeroplanes are more limited for space; however, Airbus is investing in hydrogen propulsion as the future of aviation. If they received the go-ahead today, they say they could have hydrogen-powered planes flying by 2028, and in full commercial usage by 2035. The only thing stopping us is the well-funded denialist campaign that covers up the reality of Global Warming, and the need to do something about it as soon as possible (if not sooner).

Airbus have three hydrogen-powered plane designs ready to enter the first stage of development:

- A short-range, 100-seater propeller plane similar to a Dash-8
- A medium-range single-aisle jetliner, similar to an A-320 or a Boeing 737, which could carry a few hundred passengers

- An international plane, called a 'blended wing', which looks like a flying triangle and could carry several hundred passengers on international flights.

Manufacturing, construction and buildings – from 17% emissions to virtually zero in ten years

Manufacturing and construction account for 12% of global emissions, while buildings account for 5%, making a total of 17%. Most emissions in this sector come from making two products – concrete and steel.

Concrete without burning carbon

Cement is a kind of 'glue' that sticks sand and aggregate (small rocks) together to make the final product – concrete. In 2012, we made enough concrete to build a wall 27 metres high and 27 metres wide around the Earth at the Equator (some 40,000 kilometres). Yes, we make huge quantities of concrete.

Manufacturing concrete accounts for 8% of carbon emissions, equivalent to the world's car fleet.

The traditional method of making concrete is simply to heat up calcium carbonate to make quicklime (the basis of cement) and carbon dioxide.

$$CaCO_3 + heat \rightarrow CaO + CO_2$$

The carbon dioxide is emitted from two sources: first, the colossal amount of heat needed to break down the calcium carbonate (usually provided by burning fossil fuels); and second, the carbon dioxide emitted by this specific chemical reaction. If we bypass this chemical reaction, we can avoid a huge amount of emissions.

With modern chemistry, we don't need traditional cement to glue sand and aggregate together. An effective, and in some ways superior, replacement is a family of chemicals known as geopolymer cements. Like regular cement, they join sand and aggregate and then harden. There are many different options based on different chemistry, such as slag, rock, ferro-sialate and more.

All we need to do is pass laws to phase out quicklime cement when making concrete, and to mandate using only geopolymer cements. We managed to phase out CFCs in just a few years. There's no reason why we couldn't do the same with cement made from quicklime.

Steel without burning carbon

Producing steel also accounts for about 8% of global carbon emissions.

Traditionally, to make steel, you mix iron oxide with carbon (usually high-grade coal). Then, you apply a small amount of heat (once only) to start the chemical reaction. Once the chemical reaction occurs, it gives off so much heat, you don't have to add any more.

The carbon does two things – it provides heat to make a chemical reaction happen and grabs the oxygen atom from the iron oxide, leaving behind pure iron.

$$2FeO + C \rightarrow 2Fe + CO_2 + heat$$

There's another welcome result. Not all the carbon is burned. A tiny percentage infiltrates in between the atoms of iron to make that wonderful product we call 'steel', which is far superior to iron. (Steel carries between 0.02 and 2% carbon, depending on what properties you want in the final product.)

No wonder that early steelmakers loved burning carbon with iron oxide – the carbon gave off heat and turned iron into steel.

But we can make steel without having to burn any carbon at all, because hydrogen can easily replace carbon in the chemical reaction. It would provide the heat to make the chemical reaction happen, and would also grab the oxygen atom from the iron oxide to leave behind pure iron.

$$FeO + H_2 \rightarrow Fe + H_2O + heat$$

But what about the tiny percentage of carbon (average about 0.5%) that you have to add to iron to make steel? This can easily be added to the molten iron in precise quantities. In fact, when steel is made by burning carbon with iron oxide, the resulting iron usually has too much carbon. To make modern steel, this has to be reduced to a much lower level. This extra step would thus be avoided.

If we moved to a war(like) footing we could begin the switch in months.

Buildings

Worldwide, commercial and domestic buildings take up an enormous 230,000 km^2 of land area. They could be made so much more energy-efficient, with better design, better insulation, double- and triple-glazing, natural lighting, heat sinks, courtyards, solar electricity, solar hot water, and more.

Our Australian building standards are so low in terms of energy efficiency that Europeans regularly say they have never been as cold indoors in Europe as they have been in our poorly designed and built Australian homes. In my case, when we installed double-glazing and proper insulation in our home, our annual heating bill dropped from $1,000 to $200.

Agriculture and livestock – from 11% to zero in 20–30 years

About 11% of global Greenhouse Gas emissions relate to agriculture and livestock.

There are many different ways to reduce these emissions. The first is to reduce food waste. An extraordinary one-third of all the food that's produced is wasted. In developing countries, this tends to occur in the early part of the *production* chain – growing, harvesting, storing and distribution, for example.

In wealthy countries, this waste tends to happen in the later part of the *consumption* chain. People look in the fridge, see enough food to make a meal but, instead, throw it in the bin and order a takeaway pizza.

Recent work on ruminants (cattle, sheep, etc.) has shown that adding seaweed to less than 1% of their diet can reduce methane emissions from their burps, farts and poos by 60–80%. This is worth doing because, in Australia, livestock are responsible for 60% of our agricultural emissions, and 10% of our overall domestic emissions.

There's so much more that we can do – increase plant consumption while reducing meat consumption, reducing the so-called 'food miles' by eating more locally sourced foods, regenerative farming, and making the reduction of Greenhouse Gases an actual priority.

Land use and forestry – from 6% global emissions to zero in 10–15 years

Land use and forestry (destruction and degradation) account for about 6% of our global emissions. This includes incredibly foolish acts, such as burning millennia-old rainforests to replace them with soybean crops that will survive for a few years.

It sounds ridiculously obvious to stop burning and destroying forests, and to plant more trees. Unfortunately, our currently irrational economic system makes a dead tree more valuable than a living tree. This is related to the totally improper use of that very flawed economic metric – the Gross Domestic Product, or GDP.

Fugitive emissions – from 6% of global emissions to zero in ten years

Fugitive emissions are Greenhouse Gas emissions, such as methane and carbon dioxide, caused simply by leaks. Because of the underlying assumption that the environment has no value so it's OK to pollute it, we have the wasteful and harmful situation where our society accepts this. But these account for 6% of global emissions so they're worth plugging!

GDP is not a measure of wellbeing

The inventors of the concept of GDP pointed out that it should be used only as a measure of consumption and production of goods and services – in no way should it be used as a measure of economic wellbeing. For example, the bushfires that burned 20% of all the forests in Australia in the summer of 2019–2020 also burned many houses. Those houses had to be replaced.

So the bushfires actually increased Australia's GDP!

We need a better long-term metric of a country's wellbeing.

PART 2: REVERSE GLOBAL WARMING

The second part of fixing Climate Change – which, of course, we should do simultaneously with the first part – is to reverse Global Warming. We can do this by removing carbon dioxide from the atmosphere to return levels to those recorded in the mid- to late-20th century.

One way to do this is to increase the number and size of 'carbon sinks'. In scientific or engineering terms, a sink is a location where something goes and effectively 'vanishes', hopefully never to return.

There are several methods of creating 'carbon sinks', ranging from fully biological to fully industrial – and everything in between.

Biological sinks

Biological sinks include, for example, the oceans and plants on land, which already soak up about 45% of the carbon dioxide we emit. We can increase the number of biological sinks simply by planting more trees, for starters, and cutting down less forest.

Or we can turn organic material into biochar (basically charcoal, made by heating plants in a low-oxygen environment), which can be spread onto agricultural fields to improve the soil. This charcoal is porous to water, so soaks it up. It's especially useful in floods (when it soaks up water) and droughts (when it hangs onto any available water). It can trap carbon in the soil for thousands of years. Various indigenous peoples (e.g., in the Amazon) made biochar in the past, but it is still very much underused in today's agriculture.

We could even pulverise rock into a fine dust to draw carbon dioxide from the air (if scattered into the soil) or the surrounding water (if in the ocean).

Then there's kelp farming, which sucks carbon out of the water. One way to totally offset current Greenhouse Gas emissions would be to plant kelp across about 10% of the oceans. (Hey, we have already destroyed 13 of the 17 major fisheries in the oceans, so it wouldn't hurt to put something back.) Kelp can grow 60 centimetres a day, about 30 times more than land plants. But so far, kelp harvesting is about 12 million tonnes per year, with about 75% from China.

Industrial sinks

Industrial sinks include carbon-sequestration plants (machines), which draw in the atmosphere and remove the carbon dioxide for storage. One version has been built by Climeworks in Switzerland.

To remove one year's production of carbon dioxide from the atmosphere, you'd need some 30–40 million of these small plants, each about the size of a shipping container. Luckily, these carbon-sequestration plants are much less complex to build than a plane.

It may seem that 40 million industrial plants is an impossibly large number, but it's actually less than half the number of new cars manufactured globally each year. And the modern car is far more complex to build than a carbon-sequestration plant.

If we were prepared to go without brand-new cars for half a year, we could remove a year's worth of human greenhouse emissions for that same year – and for decades afterwards. Just remember the Americans after Pearl Harbor, who ramped up production of the B-24 Liberator.

A Climeworks carbon dioxide direct air capture plant
(Photo courtesy of Climeworks)

So there are two things to realise when it comes to slashing emissions and reversing Climate Change.

First, if all the governments of the world went on to a war(like) footing (following the example of the USA after Pearl Harbor), we could start on carbon sequestration within five years. Second, this very industrial method is only one of the many potential carbon sinks available.

Ways to stop and then reverse Greenhouse Gas emissions in the atmosphere are continually being refined and updated. If you'd like to know more, here are two websites that are great starting points:

1. <u>drawdown.org</u> – A great source of research and ideas for reducing emissions. Their *Drawdown Review* is about 100 pages long and will take you an hour or so to read.

2. <u>bze.org.au</u> – Beyond Zero Emissions is packed full of factual information regarding what we can do to reach zero emissions. Cruise around this website for an hour or so, especially the section on The Million Jobs Plan.

If you spend as much time on these websites as it takes to watch a football grand final on TV, you'll see just how easy it can be to stop, and then reverse, Climate Change.

What we need are politicians who get on with doing what has to be done – just like Franklin D Roosevelt did at the beginning of World War II.

We haven't gone past the tipping point – yet.

For a full list of references, go to drkarl.com → Climate Change

Dr Karl Kruszelnicki has put his 28 years of education to good use, finding weird but true stuff about the Universe – and sharing it with fellow curious minds.

He does this via radio, TV, magazines, books, personal appearances, interpretative dance, and social media (including Twitter and TikTok).

He won the 2019 UNESCO Kalinga Prize for the Popularisation of Science, as well as an Ignobel Prize for his ground-breaking research into belly button fluff and why it is almost always blue.

This book is important because it's about our future, and how we can make it better.